Commercial Exploitation of Fisheries

Bimal Kumar Bhattacharya
for all the pleasures and pains of a very long friendship

Commercial Exploitation of Fisheries

Production, Marketing, and Finance Strategies

HRISHIKES BHATTACHARYA

OXFORD
UNIVERSITY PRESS

OXFORD
UNIVERSITY PRESS

YMCA Library Building, Jai Singh Road, New Delhi 110001

Oxford University Press is a department of the University of Oxford.
It furthers the University's objective of excellence in research, scholarship
and education by publishing worldwide in

Oxford   New York

Athens Auckland Bangkok Bogotá Buenos Aires Cape Town
Chennai Dar es Salaam Delhi Florence Hong Kong Istanbul Karachi
Kolkata Kuala Lumpur Madrid Melbourne Mexico City Mumbai Nairobi
Paris São Paulo Shanghai Singapore Taipei Tokyo Toronto Warsaw

with associated companies in Berlin Ibadan

Oxford is a registered trade mark of Oxford University Press
in the UK and in certain other countries

Published in India
By Oxford University Press, New Delhi

ISBN 019 565843 4

Typeset in Adobe Garamond by Jojy Philip, New Delhi 110 027
Printed in India at Roopak Printer, NOIDA, UP.
Published by Manzar Khan, Oxford University Press
YMCA Library Building, Jai Singh Road, New Delhi 110 001

Acknowledgements

This book is the outcome of a one and a half year study of the seafood industry. At various points during this study, extensive discussions were held with seafood exporters, trawler operators, and fishermen from various parts of the country. These interactions have enriched our understanding of the strengths and weaknesses of the Indian seafood industry.

I must thank all the seafood exporters, trawler operators, members of Fishermen's Co-operatives and Unions without whose co-operation and help the study would not have been possible. For helping in organizing the meetings and offering critical reviews at various points of the study particular mention may be made of A. Kunjumoideen, Vice President, Seafood Exporters Association of India; Ashok Nanjappa, President, Kerala region; Ravi Reddy, Vice President, Tamil Nadu region; R. Daruwala, Vice President, Maharashtra region; and also Raghunath Reddy, Rustam Irani, and Imtiaz Shroff. Thanks are also due to the many officials of banks and insurance companies who have helped the study considerably by educating us on the financial side of the industry. J.V.H. Dixitulu, Editor, Fishing Chimes, provided us with valuable information during the course of the study.

I thank K. Jose Cyriac, Chairman and G. Mohan Kumar, Director (Marketing) of the the Marine Products Export Development Authority (MPEDA); Elias Sait and Ranjit Bhattacharya, President and Vice-President, respectively, of the Seafood Exporters Association of India for their comments and suggestions. Thanks are also due to J. Ramesh, Deputy Director, MPEDA and Sandu Joseph, Secretary, SEAI for supplying the necessary data required for the study.

Thanks are also due to all the surveyors, research assistants, data analysts, and secretarial staff associated with this study. Particular mention may be made of Akshoy Singha who helped in the computer-feeding and statistical analysis of a large volume of data. My secretary, Subal Mukherjee, took tremendous care in organizing the production of the manuscript with the help of Asit Manna.

I thank the anonymous referee and the editorial staff of Oxford University Press, New Delhi for valuable suggestions, incorporation of which have enriched the book.

My wife, Gouri, has by now stopped expressing her boredom when I 'waste' midnight electricity. Orphi and Saraswat, having grown up in this parental environment, do not bother much, except for occasional whines when I cannot meet their demand for time. But all the three are always all smiles whenever my books come out in print.

HRISHIKES BHATTACHARYA

Contents

Tables

Figures

Abbreviations

BoP	balance of payments
CMIE	Centre for Monitoring Indian Economy
CMP	comprehensive market parameter
CRZ	coastal regulation zone
CV	coefficient of variation
DEPBS	Duty Exemption (Pass Book) Scheme
DSF	deep-sea fishing
ECGC	Export Credit Guarantee Organisation
EEC	European Economic Community
EEZ	exclusive economic zone
EJA	environmental impact assessment
EIC	Export Inspection Council
EU	European Union
FBE	financial break-even point
FI	financial institution
FRP	fibreglass reinforced plastic
GATT	General Agreement on Trade and Tariffs
GNP	gross national product
ICAR	Indian Council of Agricultural Research
IQF	individual quick freezing
LIBOR	london interbank offer rate

MCI	market competitive index
MEI	market expansion index
MFI	market focus indicator
MP	market parameter
MPBF	maximum permissible bank finance
MPEDA	marine products export development authority
MS	market share
MSY	maximum sustainable yield
NABARD	National Bank for Agricultural and Rural Development
NASV	net average seasonal variation
NAV	net average variation
NCAER	National Council of Applied Economic Research
NEERI	National Environmental Engineering Research Institute
NES	not elsewhere stated
NWC	net working capital
PAT	profit after tax
PC	peeled and cooked
PD	peeled deveined
PEI	product efficiency indicator
PLR	prime lending rate
PUD	peeled undeveined
SEZ	special economic zone
SD	standard deviation
SEAI	Seafood Exporters Association of India
SSI	small-scale industries
TED	turtle excluder device
UNIDO	United Nations Industrial Development Organisation
USFDA	United States Food and Drug Administration
VoP	value of production
WTO	World Trade Organization

Introduction

Unlike a regulated economy, a market economy suffers from a high degree of volatility and shocks that present many entrepreneurial challenges but, at the same time, force marginal units to the verge of extinction. The usual reaction of governments all over the world has always been to provide protection for the ostensible reason of saving these marginal units but, in effect, the industry or the sector as a whole tends to get protected by blanket protective measures. This is true particularly of sensitive sectors like agriculture and food stuffs.

Around the world, agriculture and food markets are highly protected from imports. After the successive GATT rounds, manufacturing tariffs are now at modest levels in most industrial countries and in an increasing number of middle- and low-income countries. Presently, the tariffs are between 5 and 10 per cent for manufactured goods. By contrast, agriculture and food tariffs average above 40 per cent, with tariff peaks of over 300 per cent.[1]

India is an example of consistently high tariff, with an unweighted average bound tariff of 97 per cent for unprocessed food products and 139 per cent for processed items. India, along with Pakistan, Bangladesh, and Sri Lanka, had consistently used the pressure valve of balance of payments (BOP) problems to justify the continuance of import licensing and other protective measures for agricultural goods before the GATT rounds. However, after taking on the membership of the World Trade Organisation (WTO), Pakistan, Bangladesh, and Sri Lanka have abandoned nearly all import licensing while India continues to maintain import restrictions.

Protection of agriculture is often justified in the name of food security

[1] Josling, Timothy, *Agricultural Trade Policy: Completing the Reform*, Institute for International Economics, Washington, DC, 1998.

but in reality it is more for protecting the incomes of large producers than for dealing with a food shortage. Political factors often make it difficult to remove protection. In an integrated world economy the best protection against food shortage is to have broad-based exports in order to build up a foreign exchange reserve that can be used to buy food stuffs from anywhere in the world in the event of a food shortage. But political imperatives often overshadow the economic perspective.

The market-oriented economic principle holds that in international trade a country cannot make exports if its imports are protected. A prerequisite of a national export strategy is, therefore, liberalization of imports as otherwise, countervailing or retaliatory measures are likely to be adopted by the importing countries. Despite this, agricultural protection continues and is not necessarily confined to developing countries.

In fact, the final Uruguay round of multilateral trade negotiations was held up for three years largely due to a conflict between the United States and the European Union on agricultural trade protection. Japan is well known for having one of the most protective agricultural policies of the world. Switzerland comes a close second. In Canada the tariff for butter imports has been as high as 351 per cent, and for cheese as high as 289 per cent, while other agricultural product imports have lower tariffs. Similarly, United States also has mega tariffs in dairy products and sugar. The rules and procedures formulated by the US Commerce Department in determining dumping and subsidization have a protectionist bias.[2]

Japan, the European Union countries, and the United States together account for three-fourth of India's seafood exports. These countries are well aware that like any other export oriented industry in India, seafood exports also enjoy several export subsidies and benefits, such as interest rate subsidy for loans taken by seafood export industry, capital grants and subsidy for the installation of plant and equipment, duty exemption, and import entitlements. Export profits are also exempted from income tax. In order to protect their domestic industries, these developed importing countries have been advocating the withdrawal of these subsidies in the GATT rounds and also at the World Trade Organization (WTO). So far, India has been able to withstand the pressure on the ground of BOP problems, but it may not be able to hold out for long (meat import is already liberalized).

In the post-WTO regime, protection of domestic industries is taking place through several trade and non-trade barriers. While trade barriers

[2] Baldwin, Robert E., 'Imposing Multilateral Discipline on Administered Protection', in Anne O. Krueger (ed.), The WTO as an International Organization, Oxford University Press, New Delhi, 1999, pp. 297–328.

are in the form of tariff barriers, anti-dumping measures, and general safeguards, non-trade barriers take on various shapes, including those related to environmental issues. One classic example is the passing by the United States Congress of an amendment to the Endangered Species Act, (known as Section 609) by which all shrimp exports to the United States were banned from those countries which did not employ the Turtle Excluder Device (TED) in their fishing nets to enable the endangered sea turtle to escape from the net. It is now widely acknowledged that the Section 609 amendment was meant not so much to protect the sea turtle (several other protective measures were already in place throughout the world) but to protect the ailing trawling industry of the United States against the cheaper import of shrimps from the developing nations. India, Malaysia, Pakistan, and Thailand had brought a suit against the United States in the WTO for violation of GATT obligations, as the introduction of TED increased the cost of fishing efforts and consequently the price of exports.

The second non-trade barrier comes in the form of quality standards required of imported food stuffs. These are more particularly aimed at imports from developing countries. The world is led to believe that developing countries suffer from serious quality problems in the preservation and processing of food products. During recent times, the European Union banned the import of seafood from India on the ground of quality standards. In several cases, imported goods were put under quarantine, never returned to the exporters, and eventually destroyed. Similar measures were adopted by the United States under the aegis of the Food and Drug Administration. These measures sent shock waves to the Indian seafood export industry which could not withstand the shocks owing to a low level of net worth and became sick overnight. While an internationally accepted standard of quality is always acceptable it also increases the marginal cost of exports from developing countries, which tends to protect the domestic industries of the importing countries from cheaper imports.

The burdens of anti-protective trade and non-trade barriers are always heavy on small scale operators of the export sector of a country. It has also been found that in an imperfectly competitive world trade framework, large firms can use the anti-protective laws and international rules to promote collusion between domestic and foreign firms. Under the guise of offsetting the unfair trade practices of foreign firms, domestic firms can actually bring about greater monopolistic unfairness in the domestic market.[3]

Hence, the national export strategy for agricultural and food products

[3] ibid.

should aim at improving upon the world market share in traditional exports by diversifying into newer markets and further diversifying into new exportable products in order to spread the risk emanating from monopolistic and protective trading practices of the present-day international trade regime.

The Indian seafood industry, 95 per cent of which is composed of small scale units, suffered serious setbacks in the mid-1990s after remaining a 'sunrise' industry since the late 1980s. The industry began making continuous losses with no sign of improvement. Even the top-end companies were not free from this phenomenon. At this stage the industry desired a study to be made for evolving a turnaround strategy. The present book is an outcome of this study.

At the initial stage of the study, it was found that though India is rich in marine and inland fishery resources and their contributions to the country's Gross National Product (GNP) and exports are quite significant, no comprehensive study had so far been made linking Indian fishery to the world situation. It was also observed that there is no clear-cut national strategy for the export of marine products. It, therefore, became necessary to first undertake such a study to provide the essential perspective within which the Indian seafood industry must set out its strategy for the future. This is done in Section I of the book. The major findings leading to directions of world fish trade are as follows:

(a)   There is a growing supremacy of the developing nations over the developed nations in the matter of fish production.

(b)   The trend in the share of world exports of fishery products for developing nations is on the rise on the face of a decreasing trend being observed for developed countries.

(c)   The average export price realization in fish products by the advanced developing nations has surpassed that of the developed countries. But other developing nations, like India, have not been able to obtain a better price realization, primarily owing to their technical and financial inability to carry the fish stocks.

(d)   An analysis of demand–supply gap in world fish products reveals that demand will exceed supply in the years to come.

(e)   Despite the initial enthusiasm created for prepared and preserved fish products, and the prediction that traditional fish products in fresh, frozen or, chilled form would lose the market share as a consequence of it, the direction of international trade since 1979 has proved otherwise.

Based on the above findings and further empirical investigations the following strategic business models are developed:

(a)   General Market Opportunity Indexes and Market Competitive Indexes to enable an exporting country to prioritize its strategic thrusts in different markets;
(b)   A model to evaluate market competitiveness of an exporting country in a particular fish product;
(c)   A product-specific model for strategic marketing focus to help exporters to decide on a portfolio of importing countries for a particular fish product.

In Section II, the Indian seafood industry has been analyzed. The major findings are as follows:
(a)   The structure of marine fish production in India continues to be dominantly traditional in nature despite the advent of mechanized vessels and modern fishing technology.
(b)   The rate of growth in marine fish catches in the West Coast is reaching a stagnation level due to overfishing. This is not the case in the East Coast.
(c)   The share of inland water fish landings in the total fish landings has gone up from about 29 per cent in 1980 to 42 per cent in 1997. However, the trend analysis indicates a deceleration in the rate of growth for both the fisheries. The rate of deceleration is higher in inland fisheries.
(d)   Given the existing trends in total fish production, India will face a supply shortage in the near future, even to meet the domestic consumption demand.

The findings are not very encouraging for the Indian seafood industry. What is necessary under the circumstances is first to restructure Indian marine fishing along modern lines. A scheme of such reconstruction is proposed in Chapter 7. The next task is to break the existing trends in fish production, both in the marine sector by emphasizing on the development of mechanized fleet and offshore fishing, and in inland fishing by faster development of coastal aquaculture and replacement of monoculture orientation in inland aquaculture by multiculture practices. The third proposal is to diversify deep-sea fishing (DSF) from 'shrimps only' to tuna fishing which has a huge potential in offshore waters. A comprehensive package for development of the DSF industry along with a financial restructuring plan is suggested in Chapter 8.

As mentioned earlier, units engaged in fish processing and exports have been suffering from continuous erosion of net worth during the past 4–5 years. The entire industry has virtually become sick. Presently, only about 25 per cent of the installed capacity of the industry is utilized. The procurement price equation and overhead cost structure of the industry is such that the more exports a unit makes the more losses it incurs. In the absence of net worth and banks' unwillingness to lend further, a part of the working capital loan was diverted towards a quick upgradation of plant and equipment to conform to European Union (EU) standards. This worsened the situation further. The industry is presently having a negative net working capital, and is very deep in a debt trap.

Keeping in view the problems faced by the processors and exporters, we have proposed a unique financial restructuring plan which includes the creation of a Reconstruction Fund to infuse net worth to the industry. The fund will be raised not by way of any grant but by issuance of marketable bonds of five years maturity, repayment of which will be ensured by imposition of an additional cess on all future seafood exports for five years (see Chapter 10).

We have also suggested the setting up of an autonomous quality control and monitoring agency and a Seafood Exporters Assurance Fund to ensure sustained development of the industry at par with world standards (see Chapter 10).

The book will be useful as a handbook to all those engaged in the seafood industry and the banks which finance their activities. It is also addressed to students of universities and institutes pursuing courses in agriculture and animal husbandry, and to all training and research establishments coming under the Indian Council of Agriculture Research.

I

WORLD FISHERY

1

Periods of Expansion and Decline in World Fishing
Ascendancy of Developing Nations

INTRODUCTION

It is generally held that fishery is only a small part of the economic activity of a nation in comparison to agriculture and industry. This is perhaps the reason why it still continues to be a minor concern of the public institutions. In many countries, including India, fishery does not have a separate ministry and continues to be a small part of a broad department, such as food or agriculture. What is missed, however, is its present share in the daily food platter of an individual and its linkage with other agricultural produces and the industrial economy of a nation. In any meal of carnivorous human beings, cereals constitute the bulk but the eatable animals, though small in quantity, constitute the value. Of all the meats that are consumed by people throughout the world, fish is found to be cheaper in price and softer on human biological systems.

FISHERY AS A DISTINCT ACTIVITY

The advent of commercial fishing during the last fifty years has closely linked the fishery with shipping, mechanized propelling, refrigeration, and telecommunication industries. This facilitated the process of recognizing seafood as an industry itself. However, at the social level, particularly in developing countries like India, fishery is still considered a subsidiary agricultural activity beset in the feudal perspective of fishermen, boats, and *machhwalas*. As a result, institutional attention to this field of activity has remained confined to an extension of the broad activity, namely agriculture.

Although there are more dissimilarities than similarities between agriculture and fishery, the latter is yet to be considered as a distinct segment of the economy with its unique problems and prospects. That fishery is a distinct activity, not an allied agricultural activity, has been fairly well recognized in Western countries; this is yet to be so recognized in developing economies, like India, despite the creation of several institutions concerned with fisheries development. The rules and norms applicable for agriculture are either made directly applicable to fishery or in their modified forms. Even when the seafood industry came to be recognized as such in India, no separate methodology was evolved to assess its institutional requirements. One example is that despite seafood being recognized as a distinct industry no separate norms were developed by the banking industry during the aftermath of the Tandon Committee's recommendations.[1] As a result, for the seafood industry, banks continued to apply the Maximum Permissible Bank Finance (MPBF) system without any acceptable norms for holding different current assets, while aquaculture continued to be a part of the agriculture finance department despite its well-established linkage with the seafood industry.

LOPSIDED DEVELOPMENT

Fishery development, though widely used as such in the common parlance of economic development, is somewhat ill-defined. As discussed above, sometime it is defined as an allied agricultural activity; at another time as a source of cheap but rich animal protein; sometimes as provider of employment to a large number of people; at times as another viable activity to earn precious foreign exchange for the country, and lately as a problem in environmental management with emphasis on conservation of fish resources. All these definitions came into being at different points in time and space and often to the exclusion of the other. The objectives were meant to be short run in nature. This resulted in a lopsided development of fishery. For example, the emphasis on exports created the building up of a large trawling and processing capacity in the country which now remains unutilized to a large extent. It also opened the flood-gate for the entry of a large number of players in the limited field, which caused overfishing and idle capacity. The emphasis now is on conservation which might force many a unit to go out of business. It is necessary, therefore, to conceive fishery development as an inclusive concept that encompasses all the above definitions into a single policy statement in order to prevent further lopsided development in the industry.

[1] Reserve Bank of India, *Report of the Study Group to Frame Guidelines for Follow-up of Bank Credit*, Bombay, 1975.

PERIOD OF EXPANSION

The 1950s and 1960s saw the development of world fisheries on an unprecedented scale. World fish production tripled during these two decades. This expansion was due primarily to the involvement of industrial sectors of the developed nations which innovated powered vessels, electronic fish sensing equipments, synthetic netting materials and fish specific nets, preservation and processing technology, etc. Fishing was corporatized by large organizations having financial power, information network, and worldwide marketing arrangements. These were the decades of breaking away from traditional fishing practices and development of modern skills. This reduced the marginal cost of catch and maximized the marginal revenue on an unprecedented scale. The virtual absence of governmental controls over the fishery resources fuelled this expansion.

The demonstration effect of fisheries development in the western hemisphere had a fallout effect in developing nations, which line the shores of most of the oceans, to vigorously move for a new Law of the Sea in the 1970s, which saw the declaration of a 200-mile Exclusive Economic Zone (EEZ) for coastal countries in 1975. More than 95 per cent of the world's conventional marine fishery resources are concentrated in these EEZs. New hopes emerged in developing coastal nations who now wanted to have a larger share in the use of fishery resources, both in respect of domestic consumption and exports. These expectations were reinforced when the Food and Agriculture Organisation (FAO) of United Nations declared that a revolution had occurred in the potential of fisheries to contribute to a new international order, and the FAO intended to take the lead by helping the developing countries to secure their rightful place in world fisheries.[2] This was followed up by several studies and consulting activities by FAO to equip the developing countries with modern methods of fishing, including scientific methods of estimating fishery resources and their conservation.

FISHERY EXPANSION IN THE DEVELOPING WORLD

All the above developments led to unprecedented growth in the fish catches of the developing countries. The share of the developing countries in the world total of nominal catches increased from 43.23 per cent in 1975 to 62.92 per cent in 1992. During the same period the share of the developed countries fell from 56.77 per cent in 1975 to 37.08 per cent in 1992, as will be evident from Table 1.1.

---

[2] 'Fisheries Development in the 1980s', Food and Agriculture Organisation, Rome, 1984.

TABLE 1.1
Percentage share of world nominal catches of fish,
crustaceans, molluscs, etc.

Year	% share of developing countries	% share of developed countries
1975	43.23	56.77
1976	43.98	56.02
1977	44.16	55.84
1978	46.80	53.20
1979	47.66	52.34
1980	46.83	53.17
1981	48.32	51.68
1982	48.51	51.49
1983	47.19	52.81
1984	48.47	51.53
1985	50.76	49.24
1986	52.54	47.46
1987	51.91	48.09
1988	53.56	46.44
1989	56.18	43.82
1990	57.97	42.03
1991	60.65	39.35
1992	62.92	37.08

Source: Calculated from *Fishery Statistics Year Book*, Food and Agriculture Organisation, various issues.

The cumulative rate of growth of catches in the developing countries during 1975–92 had been an impressive 4.45 per cent per annum while the same was almost zero for the developed nations during the same period. The stagnant growth rate in the developed countries can be attributed to overfishing in the 1960s, caused by unbridled expansion of fishing activity spearheaded by the operation of about 25,000 huge fishing vessels measuring 60 metres and above with Horse Power capacity between 500 and 2000, and in some cases even above.

SUPREMACY OF THE DEVELOPING NATIONS

Table 1.1 clearly reveals the growing supremacy of developing countries over the developed nations of the world in the matter of fish production. It also indicates that the developed countries have become dependent on the developing nations for the supply of fish and will continue to remain so for the years to come. This view is supported by the increasing share of the developing nations in the total exports of the world, as will be evident from Table 1.2.

TABLE 1.2
Percentage share of world exports in terms of US dollars

Year	% share of developing countries	% share of developed countries
1975	36.96	63.04
1976	34.62	65.38
1977	35.94	64.06
1978	36.86	63.14
1979	37.44	62.56
1980	37.40	62.60
1981	39.30	60.70
1982	40.76	59.24
1983	40.40	59.60
1984	44.21	55.79
1985	44.01	55.99
1986	45.38	54.62
1987	45.51	54.49
1988	46.14	53.86
1989	46.60	53.40
1990	44.00	56.00
1991	45.90	54.10
1992	46.92	53.58
1993	48.98	51.02

Source: Fishery Statistics: Commodities Year Book, Food and Agriculture Organisation, Rome, various issues.

There are clear trends that the percentage share of the developed nations in world exports is falling while that of the developing nations is on the rise. In 1993 the share of the developing nations had gone up to nearly 50 per cent of world exports in US dollar terms, which is likely to increase in future.

DEPLETION OF OCEANIC RESOURCES

The major fishing areas of most of the developed countries are along the Atlantic Ocean which is also the major source of world fish production, second only to the Pacific Ocean. But the Atlantic's share of World oceanic catches is falling continuously. In 1975 it had about 41 per cent share in world nominal catches, which came down to nearly 25 per cent in 1996. Except the Western Central and South West Zones of the Atlantic (mostly lined by developing nations), all other zones are experiencing declining trend in their catches due to overfishing. The depletion was high in the North East Atlantic. But due to conservation measures this Atlantic Zone is presently enjoying a rebounding effect.

The Pacific Ocean, which is largest among all the oceans, also supplies the largest quantity of fish to the world. Unlike the Atlantic, its share of world catches increased from 52.82 per cent in 1975 to 65.44 per cent in 1996. This ocean is also suffering from overexploitation, particularly the North-west Pacific which is fished primarily by the United States and Canada. The South East Zone, belonging mainly to the developing nations of South America, has suffered comparatively less on account of overfishing. Presently, the South American nations are also among the major suppliers of fish to United States.

The Indian Ocean, whose coastal areas are mostly peopled by developing countries, is about 79 lakh sq. km. short of the Atlantic but it has a share of only 9.23 per cent of all oceanic catches in 1996 (this increased from 5.38 per cent in 1975). However, the growth rate in catches of the Indian Ocean is highest among all the oceans. There is substantial variation in catch characteristic between the Western and Eastern Zones of the Indian Ocean. The Western Zone had a share of 65.39 per cent of total catches of Indian Ocean in 1975, which has declined to 51.10 per cent in 1996. In contrast, the Eastern Zone had a share of 34.61 per cent in 1975, which increased to 48.90 per cent in 1996. There are perceptible signs of overfishing in the Western Zone.

The Mediterranean Sea and Black Sea are major fishing areas of the developed coastal states of Europe. These together had a share of 2.06 per cent of the total world marine catches in 1975, which has declined to 1.72 per cent in 1996. These two areas have also suffered from overfishing. But due to the adoption of conservation measures they are presently enjoying a rebounding effect since 1992.

A graphical presentation of fish production of the oceanic zones is given in Figs. 1.1–1.4. Table 1.3 summarizes the characteristics of the three oceans and their fishery resources.

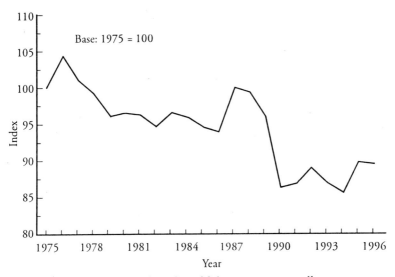

*Fig. 1.1:   Nominal catches of fish, crustaceans, molluscs, etc.
in the Atlantic Ocean*

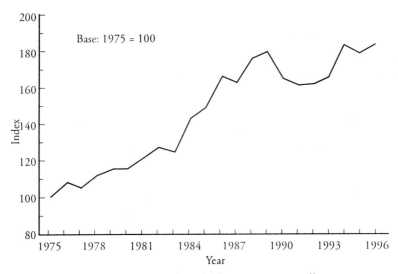

*Fig. 1.2:   Nominal catches of fish, crustaceans, molluscs, etc.
in the Pacific Ocean*

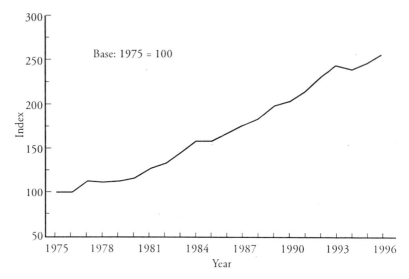

*Fig. 1.3:   Nominal catches of fish, crustaceans, molluscs, etc.
in the Indian Ocean*

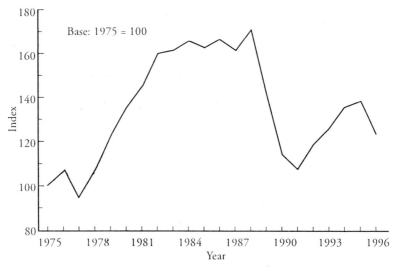

*Fig. 1.4:   Nominal catches of fish, crustaceans, molluscs, etc.
in the Mediterranean Sea and the Black Sea*

TABLE 1.3
Characteristics of oceanic fish resources

	Atlantic Ocean	Pacific Ocean	Indian Ocean
Area (in thousand sq. km.)	80,419	166,959	72,520
Percentage share of catches			
1975	41.80	52.82	5.38
1996	25.33	65.44	9.23
Cumulative growth rate of catches (% p.a.)			
1975–96	(–) 0.53	2.93	4.53
1985–96	(–) 0.49	1.89	4.38
Mean growth rate (year to year in %)	(–) 0.47	3.06	4.59
Fluctuations in catches in (CV %)	(–) 717.22	174.39	75.07
Growth characteristics	Decreasing at an increasing rate	Increasing at a decreasing rate	Increasing at a decreasing rate
Projected level of catches in 2005 (thousand metric tons)	19,648	69,417	10,053
R^2 (Significant at 1% level)	0.67	0.89	0.98

Note: Derivation of the above parameters is based on time series data given in the Annexe 1.1.

The findings in Table 1.3 suggest that the performance of the Indian Ocean in all the growth parameters is best among all the oceans. It has the highest growth rate in catches, with comparatively low fluctuations. The share of oceanic catches of Indian Ocean has increased by 71.56 per cent during 1975–96. The Atlantic Ocean has fared poorly in all the growth parameters. It has lost its share of catches by 39.40 per cent during the period under study. The growth characteristics of all the oceans, however, indicate overfishing as a general phenomenon for all the oceans.

CONCLUSION

The developed countries as a whole have lost the share of catches to the developing world by about 35 per cent during 1975–96. The trend suggests that this loss will be higher in the years to come. Developing nations located in the coastal areas of the Pacific and the Indian Ocean are going to be the major suppliers of fish to the world. This conclusion is also supported by the increasing share of fish exports by the developing nations. They have gained over developed countries by about 20 per cent during the period under study and, presently, they control about 50 per cent of the world export of fish products. All these findings put the developing nations at a tremendous comparative advantage over the developed countries, who will be increasingly dependent upon the developing nations for their future supply of fish products. Those countries in the developing world who had realized this comparative advantage early have increased their bargaining power for better price realization by increasing both physical and financial carrying capacity of stocks of fish. This shall be examined further in the next chapter.

ANNEXE 1.1
Trend in nominal catches of different oceanic fishing zones

(in thousand metric tons)

Year	Atlantic Ocean	Pacific Ocean	Indian Ocean	Mediterranean and Black Sea
1975	24,188	30,561	3112	1215
1976	25,258	32,912	3143	1307
1977	24,440	32,072	3470	1153
1978	24,015	34,119	3467	1319
1979	23,244	35,122	3480	1494
1980	23,322	35,337	3554	1642
1981	23,262	37,102	3931	1760
1982	22,876	38,766	4112	1945
1983	23,326	38,095	4501	1961
1984	23,179	43,573	4869	2015
1985	22,860	45,531	4922	1979
1986	22,709	50,684	5175	2022
1987	24,147	49,709	5401	1961
1988	24,017	53,446	5667	2079
1989	23,180	54,878	6110	1716
1990	20,872	50,335	6279	1385
1991	20,975	49,368	6631	1309
1992	21,531	49,513	7102	1440
1993	21,038	50,424	7542	1525
1994	20,724	55,930	7388	1644
1995	21,691	54,509	7616	1675
1996	21,656	55,952	7889	1496

Cumulative growth rate (% p.a.)

1975–96	(–) 0.53	2.92	4.53	1.00
1985–96	(–) 0.49	1.89	4.38	(–) 2.51

2

Price Realization of Fish Products

INTRODUCTION

Price realization in fish products by the various exporting countries depend on the periodic abundance or shortfall in their catches *vis-à-vis* the world supply, species and sizes of individual fish captured and their demand in the world market, the standard of hygiene at all levels of fish production, and above all the physical (storage) and financial capacity to carry the product stocks.

In Tables 2.1 and 2.2, the average price realization of fish (fresh, chilled, frozen, dried, salted, or smoked) and crustaceans (and molluscs) have been calculated for the major developing and developed nations of the world. Developing countries are divided into two groups. The Group A countries are the advanced developing countries; the Group B countries are not so advanced.

TABLE 2.1

Average per metric ton price realization
of fish and crustaceans by major exporters of developing countries

					(in thousand US dollars)
	1993	1994	1995	1996	Average
GROUP A					
Brazil					
Fish	1.61	1.49	2.22	2.17	1.87
Crustaceans	8.87	11.69	15.43	15.23	12.80

(Contd.)

Table 2.1 (contd.)

	1993	1994	1995	1996	Average
Mexico					
Fish	1.61	2.81	1.28	1.37	1.77
Crustaceans	12.66	10.44	9.27	6.42	9.70
Chile					
Fish	3.45	3.57	4.00	3.57	3.65
Crustaceans	5.21	6.52	7.70	8.79	7.06
Indonesia					
Fish	1.33	1.35	1.39	1.37	1.36
Crustaceans	7.93	8.98	9.83	9.24	9.00
Korea(R)					
Fish	2.94	3.57	3.45	3.12	3.27
Crustaceans	7.56	7.58	4.29	4.13	5.89
Philippines					
Fish	1.71	1.73	2.06	2.11	1.90
Crustaceans	7.40	8.02	8.43	5.52	7.34
Singapore					
Fish	3.25	3.02	3.09	3.48	3.21
Crustaceans	5.19	5.23	6.35	6.17	5.74
Thailand					
Fish	1.10	1.21	1.33	1.29	1.23
Crustaceans	8.20	8.69	9.78	8.70	8.84
Average					
Fish	2.12	2.34	2.35	2.31	2.28
Crustaceans	7.88	8.39	8.89	8.02	8.30
GROUP B					
Argentina					
Fish	1.25	1.23	1.45	1.37	1.33
Crustaceans	1.50	1.45	1.63	1.62	1.55
Bangladesh					
Fish	3.84	2.32	1.67	3.70	2.88
Crustaceans	6.80	8.06	9.94	8.17	8.24
China					
Fish	2.22	2.78	3.12	2.56	2.67
Crustaceans	3.17	3.57	3.79	2.90	3.36

(Contd.)

Table 2.1 (contd.)

	1993	1994	1995	1996	Average
India					
Fish	1.10	1.19	1.18	1.12	1.15
Crustaceans	4.70	5.58	5.45	4.66	5.10
Malaysia					
Fish	1.33	1.03	0.90	0.98	1.06
Crustaceans	1.25	1.46	1.64	1.35	1.42
Morocco					
Fish	2.50	2.50	3.12	3.12	2.81
Crustaceans	2.97	3.53	4.65	5.11	4.07
Pakistan					
Fish	1.69	1.31	1.14	1.18	1.33
Crustaceans	3.38	3.96	4.42	3.94	3.93
Average					
Fish	1.99	1.77	1.80	2.00	1.89
Crustaceans	3.40	3.94	4.50	3.96	3.95

Source: Fishery Statistics: Commodities Year Book, Food and Agriculture Organisation, Rome, various issues.

TABLE 2.2

Average per metric ton price realization of fish and crustaceans of major exporters of developed countries

(in thousand US dollars)

	1993	1994	1995	1996	Average
Australia					
Fish	5.80	5.00	6.07	6.29	5.79
Crustaceans	15.88	16.00	16.08	15.70	15.92
Canada					
Fish	3.28	3.56	3.70	3.59	3.53
Crustaceans	7.66	8.02	8.83	7.76	8.07
Denmark					
Fish	3.20	3.45	3.62	3.31	3.40
Crustaceans	3.77	4.22	4.48	4.33	4.20

(Contd.)

Table 2.2 (contd.)

	1993	1994	1995	1996	Average
France					
Fish	1.95	1.98	2.23	2.15	2.08
Crustaceans	4.03	4.63	5.39	4.07	4.53
Hong Kong					
Fish	1.88	2.17	2.30	2.75	2.28
Crustaceans	4.16	4.74	4.54	4.86	4.58
Japan					
Fish	1.32	1.65	1.97	1.89	1.71
Crustaceans	4.54	5.63	4.78	2.14	4.27
Norway					
Fish	1.97	2.02	2.14	2.09	2.06
Crustaceans	5.76	5.02	6.80	5.65	5.81
United Kingdom					
Fish	1.29	1.54	1.62	1.72	1.54
Crustaceans	4.71	5.10	5.20	4.99	5.00
United States					
Fish	2.82	2.90	3.10	2.76	2.90
Crustaceans	6.42	6.15	6.20	5.48	6.06
Belgium					
Fish	3.59	4.13	4.19	4.32	4.06
Crustaceans	6.62	7.23	7.87	6.84	7.14
Average					
Fish	2.71	2.84	3.09	3.09	2.94
Crustaceans	6.35	6.67	7.02	6.18	6.56

Source: As in Table 2.1.

COMPARATIVE ANALYSIS BETWEEN DEVELOPED AND DEVELOPING COUNTRIES

It may be seen that the average price realization of crustaceans by the advanced developing nations in Group A have surpassed that of developed countries. In the case of fish items, they are close on the heel. This is despite the general belief that developed nations produce (catch) high value fishes in the Atlantic and Pacific Oceans. Group A countries are able to do this by virtue of their modern storage facilities, which can carry stocks in a good hygienic condition for a longer period than by traditional facilities, and the financial capacity matching with it to obtain better prices in the world market. Among the developing nations price realization in crustaceans is highest for Brazil, followed by Mexico.

COMPETITIVE ANALYSIS AMONG EMERGING COUNTRIES OF ASIA

THAILAND

Indonesia and Thailand closely follow Chile in the price realization of crustaceans. About 30 per cent of Thailand's exports of crustaceans go to the United States and Canada. Thailand's shrimp export has been experiencing an annual growth rate of nearly 22 per cent since 1989 in quantitative terms. This is despite the fact that the country has been experiencing a declining trend in marine shrimp landings since 1982.[1]

Aquaculture

Realizing that the maximum sustainable yield in marine shrimp production has already been overexploited, Thailand quickly developed an alternative strategy to boost up the aquaculture production of shrimp. Through concerted policy thrusts the share of cultured production of shrimp went up from 6.1 per cent in 1982 to 72.7 per cent in 1994. Intensive aquaculture, with an yield of about 6 tonnes per hectare, now occupies more than 50 per cent of the hectarage, contributing more than 96 per cent of the total aquaculture production of Thailand. The Department of Fisheries of Thailand has recently undertaken a large construction work for establishing sea water irrigation systems in the principal shrimp production areas of the state in order to improve water exchange, with a plan to excavate sand bars at the river mouths to allow better exchange of water.[2]

Modernization

Thailand's next strategy to boost exports was to rapidly modernize its processing and storage facilities. More than 50 new processing plants were established between 1990 and 1993 with all modern facilities conforming to international standards, including post-harvest processing facilities. Simultaneously, emphasis was placed on infrastructure development. Thailand now boasts of a well-developed infrastructure in terms of modern roads, telecommunication, and power. But landing facilities for marine

---

[1] *Foreign Trade Statistics of Thailand*, 1993–94, Ministry of Finance, Government of Thailand, Thailand, 1995.

[2] Rungarai, Tokrisna, 'Aquaculture in Thailand', Aquaculture in Asia: Prospects of the 1990s, APO Symposium on Aquaculture, 1992.

catches are not yet satisfactory. In all the 16 major fishing ports which account for nearly 80 per cent of landings, there are problems of unloading and preservation. The Fish Marketing Organization, the public sector arm of the Ministry of Agriculture and Cooperatives, is taking steps to modernize the fish handling systems at landing centres and fish markets.

Incentives

Although Thailand has withdrawn all fiscal incentives (except rebate on customs duty) since 1993, financial support to the seafood export industry continues. Fifty per cent of the export credit is made available to the exporters by the Board of Trade at an interest rate varying between 3 and 4 per cent. The remaining 50 per cent is made available by commercial banks at a commercial rate of interest.

INDONESIA

Indonesia's price realization of crustaceans is best among all the developing nations of Asia, though it suffered a decline during 1988–92. Shrimp is the major item of marine exports of Indonesia, though its share among other fish products is declining. But unlike Thailand, marine shrimps account for more than 50 per cent of the total shrimp production. Indonesia is yet to acquire sufficient technical and managerial skills needed for shrimp cultivation.[3] Marine shrimp fishery is growing at about 5 per cent per annum. Indonesia's major export markets are Japan, South East Asia, and the United States. In Japan, Indonesia is the market leader in shrimps.

Modernization

Indonesia has well-developed processing and storage facilities (though on-board storage facilities are limited). There are about 180 modern freezing plants of international standard and 22 canning factories.[4] But infrastructural support for fish production is lacking. The present fishing harbours and landing centres are not sufficient to handle large volumes. The government has embarked on a long-term plan to construct fishing harbours and landing centres with modern facilities in the east and south-east of Indonesia. There is also a plan to construct an irrigation canal to boost up aquaculture of shrimp with the financial assistance of the World Bank.

As in Thailand, at present there are not many fiscal incentives available

[3] Supardan, Ali, 'Aquaculture Development in Indonesia', Aquaculture in Asia, APO Symposium in Aquaculture, 1992.
[4] 'Indonesia: Industrial Growth and Diversification', Industrial Development Review Series, UNIDO, 1993.

to the seafood exporters in Indonesia, though export credits are available at subsidized rates.

REPUBLIC OF KOREA

In the Republic of Korea (Korea(R)) the price realization for crustaceans is lower than in the other major Asian neighbours in Group A countries but the price realization in fish products is highest among all of them. Korea (R) has specialized in distant water fishing and fishes in all the oceans of the world by agreement with several countries. Deep-sea fishing enables it to capture high value fishes like big eye tuna, croaker, yellowfin tuna, etc. The total tuna production of Korea(R) is also quite high, at around 170 thousand metric tonnes, though production from deep-sea fishing is declining over the years.[5] In order to conserve and strengthen marine resources, Korea(R) has prohibited fishing operations during breeding seasons and imposed restrictions on fish catches below a specified size.

Aquaculture

Korea(R) does not have much scope for inland shrimp culture, particularly black tiger shrimps of high value varieties, because of unsuitable climatic conditions. The major source of shrimp landings is marine, though high value shrimp landings from distant water is declining. The contribution of aquaculture is negligible. All these resulted in a low value realization of the crustacean group of products.

Nearly 50 per cent of Korea(R)'s export presently goes to Japan in quantitative terms. America and Europe account for about 8 per cent each. However, Korea(R) is unable to maintain its market share in the countries of her exports, as we shall see later.

The country is also an importer of shrimps; imports are often more than the export of shrimp. The bulk of these imports are re-exported at a value higher than that of imports.

Production Facilities

Though Korea(R) has excellent frozen capacity and processing units for specialty products, it faces the problem of shortage of skilled labour and consequent high wage cost. In order to counteract the problem the Korean businesses have adopted the strategy of relocating their production facilities in other countries of South East Asia where labour cost is comparatively cheap.

[5] *Deep Sea Fisheries in Korea*, Korea Deep Sea Fisheries Association, April, 1994.

Incentives

The country offers several tax incentives and import duty concessions to the seafood industries, particularly those located in the Free Export Zones.[6]

REALIZATION OF COMPETITIVE ADVANTAGE

The above brief analyses of some of India's neighbouring fish exporting countries indicates that the existence of competitive advantage does not always result in a better price realization. The countries analysed here have shown that in order to encash the competitive advantage in the seafood industry a country has to have both technical and financial power to carry her stocks of fish products for better price realization. Barring a very small number, developing nations falling under Group B, which also includes India, lack these two powers. The average price realization of this group of countries, both in fish and crustaceans, is much lower than that of the developed nations, whereas the latter nations have been surpassed by a wide margin by the advanced developing nations.

CONCLUSION

Table 2.1 and 2.2 indicate that even in matters of price realization of fish and crustaceans, the developed fish exporting countries have lost their competitive advantage. Barring Australia and Canada, all other countries have low price realization. If we exclude these two countries from the list, the average price realization of the remaining eight developed countries will be close to those of the Group B developing countries. It is unlikely that the developed fish exporting countries will be able to make a turnaround in price realization as they have lost control over catches. Some of the Group B developing countries are major producers of fish (India for example). But this advantage is not exploited due to the lack of technical and financial power to process, preserve, and hold fish stocks for better price realization.

---

[6] *Reform Plans for Improvement in Foreign Investment Environment,* Ministry of Finance, Republic of Korea, 1994.

3

Global Fish Production
Market Opportunities and Competition

INTRODUCTION

There has been a general decline in the rate of growth of global marine fish production. The period of expansion during the 1950s and 1960s was to a large extent uncontrolled. This resulted in overfishing and consequent stagnancy and/or depletion of marine fishery resources. The impact of this was and is being felt in the subsequent decades in the general fall in the rate of growth of nominal catches of all types of fishes. Also, some valuable fish species have become near-extinct, besides the disturbance of the undersea aquatic balance, which we shall discuss in Chapter 4.

The world nominal catches of fish, crustaceans, and molluscs has slowed down from 6 and 7 per cent annual rate of growth during the 1950s and 1960s respectively to a mere 2.62 per cent per annum during 1975–96. When we consider the more recent period, 1985–96, the annual growth rate has declined to 2.03 per cent.

DEVELOPED COUNTRIES

All the major producers of fish from the developed countries, except United States, Norway, and Iceland, have negative growth rates in fish production during 1975–96. However, in the recent period (1987–96) the positive growth rate of the United States (2.73 per cent per annum) turned to a negative growth rate of 6.70 per cent per annum while Norway's position improved from 0.28 per cent per annum to 3.76 per cent per annum during the same period. Norway's total catches started declining from 1985 but rebounded back from 1991 due to the adoption of conservation measures. We have not been able to obtain consistent series of data for USSR since

1991 owing to the dismantling of the Republic. However, an analysis of data from 1975 to 1990 indicate an overall cumulative growth rate of 0.28 per cent, which improved to 1.12 per cent per annum from 1987.

The deceleration in growth rate is highest in the case of Japan. Its share of world catches has also declined to 5.27 per cent in 1996 from 15.06 per cent in 1975. The process of deceleration began in 1987 with no sign of abatement. Considering the fact that the per capita annual human consumption of fish in Japan is among the highest in the world (69.9 kg during 1993–5, though reduced from 72.1 kg during 1987–89),[1] her dependence on imports will be substantial in the years to come, even though there is a cut down in per capita consumption by 3.05 per cent despite a rise in population by 1.78 per cent between the two periods. Presently, Japan is consuming about 9.75 per cent of the world fish production, including non-food uses, of which about 32 per cent is coming from net imports. The consumption is likely to increase in future years with the rise in the Gross National Product (GNP). However, assuming that the same trends continue, Japan's human consumption (excluding non-food uses) of fish will be about 8577 thousand metric tons by 2005 (8620 metric tons by 2002).

The United States has been experiencing a decline in catches during recent years. It maintained its share of world catches of more than 5 per cent for a long period (1980–94), after which its share began declining, and at the end of 1996 reverted back to its market share in 1975. The annual average per capita human consumption of fish in the United States has been 21.9 kg during 1993–5 (increased from 21.3 kg during 1987–9). Presently, the United States is consuming about 6.15 per cent of the world fish production, including non-food uses, of which nearly 15 per cent comes from net imports. Between 1987 and 1995 the population of the United States rose by 7.55 per cent and the per capita consumption of fish also rose by about 2.82 per cent during the same period. Assuming that the same trends continue, the consumption of fish in the United States will rise to a level of about 6570 thousand metric tons by 2005 (about 6200 thousand metric tons by 2002).

Norway is the largest producer of fish among the market economies of Europe. It is also the only country among the major fish producers of the developed world which has registered a positive cumulative growth rate during 1987–96 for reasons mentioned earlier. The per capita consumption of fish in Norway is also very high. It was 47.5 kg during 1993–5. However, Norway consumes only about 0.78 per cent of the world fish production. It is a small country with a population of about 43 lakhs which is growing at

---

[1] *Fishery Statistics: Commodities Year Book*, Food and Agriculture Organisation, Rome.

a rate of around 0.30 per cent per annum. The country is a net exporter of fish products. During 1993–5 Norway's net export of fish products was about 53 per cent of its total production.

Denmark comes after Norway in respect of her share of world fish production, though the share has fallen from 2.69 per cent in 1975 to 1.49 per cent in 1996. Unlike Norway, Denmark's production has stagnated over the years. The per capita annual consumption of fish is 20 kg which is only about 0.09 per cent of the world fish production. Denmark is also a small country. She is a net exporter of fish products, though they constitute only about 13.6 per cent of total fish production.

Iceland is another country in the developed world which has a positive growth rate in fish production, though its cumulative growth rate has fallen during recent years with a marginal reduction in the share of world production. Her per capita consumption of fishery products is 91.2 kg (1993–5), which is the highest among the developed nations and second in the world (the first being Maldives with 139.8 kg per capita consumption). Iceland is the second smallest country of the developed world, with a population of only about 2.7 lakh. The country is a net exporter of fish products which is about 40 per cent of total fish production.

NET EXPORTERS

The countries of the developed world who are net exporters of fish and fishery products during 1993–5 are given in Table 3.1.

TABLE-3.1
Average net exporters of fish and fishery products in the developed world

(average of 1993–5)

Country	Average net export as % of total production	Average net export as % of world export
Canada	28.96	1.43
New Zealand	82.86	2.02
Denmark	13.62	1.18
Ireland	74.64	1.21
The Netherlands	52.65	1.29
Faeroe IS	68.70	0.42
Iceland	69.08	0.84

(Contd.)

Table 3.1 (contd.)

Country	Average net export as % of total production	Average net export as % of world export
Norway	52.50	3.02
South Africa	12.41	6.43
Sweden	23.12	0.32
Total	39.86	18.16

Notes: Percentages are calculated on weights in kg; production and next export refer to fish, crustaceans and molluscs including all aquatic organisms except whales and sea weeds.

Source: Basic data are taken from Appendix I of *Fishery Statistics, FAO Yearbook, 1996*.

Table 3.1 indicates that of the 26 countries of the developed world (excluding transition economies) considered here, 10 countries have net exports that constitute 40 per cent of their average total production of fish and fishery products and about 18 per cent of average world exports; significant among them is South Africa whose net export is lowest as a percentage of total production but highest as a percentage of world exports. The remaining 16 countries are net importers of fish and fishery products, as shown in Table 3.2.

TABLE 3.2

Average net importers of fish and fishery products in the developed world

(average of 1993–5)

Country	Average net import as % of consumption	Average net import as % of world imports
United States	14.66	4.63
Australia	36.10	0.60
Austria	92.94	0.36
Belgium	84.75	0.92
Finland	29.70	0.35
France	50.30	3.89
Germany	71.48	3.40
Greece	20.38	0.24
Italy	74.10	4.24
Portugal	54.58	1.53
Spain	36.90	3.49

(Contd.)

Table 3.2 (contd.)

Country	Average net import as % of consumption	Average net import as % of world imports
United Kingdom	13.07	0.69
Malta	75.00	0.03
Switzerland	96.77	0.41
Israel	82.08	0.40
Japan	32.18	16.1
Total	33.34	41.32

Notes: Percentages are calculated on weights in kg; consumption includes non-food uses; consumption and net import refer to fish, crustaceans, and molluscs including all aquatic organisms except whales and seaweed.
Source: As in Table 3.1.

NET IMPORTERS

Table 3.2 indicates that the average net imports of 16 developed countries is more than 33 per cent of their consumption of fish products and constitute about 41 per cent of world imports. Significant among them is Japan whose net import as a percentage of consumption is about 32. Japan is also the largest net importer of fish products among all developed countries. Italy's net import is 74 per cent of her total consumption of fish products. Italy's share of world imports at 4.24 per cent is also quite significant. For the United States the net import as a percentage of consumption is low at 14.66, but her share in world import is significant at 4.63 per cent.

The average net import of fish products by the developed world as a whole (excluding transition economies) is 14.76 per cent of consumption, including non-food uses.

TRANSITION ECONOMIES

The transition economies, which were earlier listed as part of the developed world, are all net importers of fish products, except the former USSR areas, as shown in Table 3.3.

The only country that is significant among the net importers of the transition economies is Poland, its net import being about 32 per cent of consumption and about 1 per cent of world imports.

The net export of the areas belonging to the former USSR is about 25.82 per cent of their fish production, which constitutes about 6.04 per cent of world exports of fish products. The average share of the former USSR areas in world fish production was 4.5 per cent during 1993–5.

The above analyses indicate that of all the developed nations (including those of transition economies) 23 countries are net importers of fish products, which together constitute about 32.56 per cent of total consumption and 42.52 per cent of world imports. The remaining 11 developed countries, including the former USSR, are net exporters of fish and fishery products, which constitute about 36 per cent of their total fishery production and 24.20 per cent of world fish exports.

TABLE 3.3

Net import of fish and fishery products by developed transition economies

(average of 1993–5)

Country	Average net import as % of consumption	Average net import as % of world imports
Albania	22.55	0.003
Bulgaria	15.10	0.017
Czechoslovakia	50.00	0.112
Hungary	42.70	0.080
Poland	31.64	0.945
Romania	19.16	0.038
Yugoslavia	4.70	0.001
Total	31.88	1.196

Source: As in Table 3.1.

MARKET OPPORTUNITY INDEX

In the fish and fishery products market of the developed world, opportunities are provided by those countries whose import content of consumption as well as the share of world import are high. These two percentage figures, as appearing in Table 3.2 and 3.3, are therefore, multiplied to develop a weighted index of market opportunities for the purpose of ranking these countries in descending order of opportunities provided by them. This is done in Table 3.4.

Japan ranks first in providing market opportunities for fish products because her dependence on import is very high—32.18 per cent of consumption—as well as the volume of import—net import being 16.14 per cent of the world import. The United States ranks eighth because her dependence on import is comparatively lower—14.66 per cent of consumption—as well as the volume—net import being 4.63 per cent of world imports. Italy ranks second because her dependence on import is very high though volume-wise she is close to the United States. Similar is the case with Germany. Malta ranks lowest in the list; Malta has a high dependence on imports, but volume-wise the import is very low and, therefore, Malta does not provide much market opportunities.

TABLE 3.4

Country ranking of the developed world by Market Opportunity Index

Rank	Country	Market Opportunity Index
1	Japan	519.38
2	Italy	314.18
3	Germany	243.03
4	France	195.67
5	Spain	128.78
6	Portugal	83.50
7	Belgium	77.97
8	United States	67.87
9	Switzerland	39.68
10	Transition economies	38.12
	(Group of 7 countries as in Table 3.3)	
11	Austria	33.46
12	Israel	32.83
13	Australia	21.66
14	Finland	10.40
15	United Kingdom	9.01
16	Greece	4.89
17	Malta	2.25

COMPETITIVE INDEX

A country which has a low level of domestic consumption denoted by net export as a percentage of production, and high volume of export denoted by net export as a percentage of world export, is one which offers stiff competition in the market of fish and fishery products. We have accordingly ranked the net exporting countries of the developed world in accordance with the competitive index drawn up by multiplying these two percentages as appearing in Table 3.1 and as calculated for the erstwhile USSR areas. These ranking are given in Table 3.5.

New Zealand ranks highest in the Competitive Index because her domestic consumption is low, which permits her to export a large part of her production. Volume-wise, the export is also significant, net export being

2.02 per cent of the world export. South Africa, despite having the largest share of the world export among the net exporters of the developed countries, ranks only fifth in competitiveness because her domestic consumption is high, permitting her to export (net) only 12.41 per cent of production. That is, South Africa has much less competitive leverage than, say, New Zealand or Norway.

Table 3.5
Country ranking of the developed world by Competitive Index

Rank	Country	Competitive Index
1	New Zealand	167.38
2	Norway	158.55
3	Former USSR	155.95
4	Ireland	90.31
5	South Africa	79.80
6	The Netherlands	67.91
7	Iceland	58.03
8	Canada	41.41
9	Faeroe IS	28.85
10	Denmark	16.07
11	Sweden	7.40

DEVELOPING COUNTRIES

CHINA: THE LEADING PRODUCER

All the major fish producing countries of the developing world have positive growth rates in fish production.[2] China has emerged as the number one country, displacing Japan, which at one time was the largest producer. China's share of world nominal catches, which was 6.46 per cent in 1975, went up to 18.50 per cent in 1996. The other fish producing countries of both developing and developed worlds are far behind China.

China is a vast country with long coast lines and huge internal water resources, which is the major strength of China's fishery industry. The potential inland water area is around 17.47 million hectares of which nearly 12.8 million hectares comprise lakes and rivers. This has enabled the country to withstand the falling trend in her marine catches by quickly moving towards a strategy of faster development of aquaculture.

[2] Countries considered here are China, Malaysia, Philippines, Korea (DPR), Korea (R), Vietnam, India, Indonesia, Thailand, Mexico, Argentina, Brazil, Peru, and Chile.

For the entire period of 1975–96, China's cumulative growth rate in catches has been 7.89 per cent per annum. However, since 1987 the rate of growth accelerated which resulted in a cumulative growth rate of 8.32 per cent per annum—much larger than the cumulative growth rate in the world production of fishery products.

The phenomenal growth rate in China's nominal catches is the result of the liberalization policies of the government and placing fish production and export as top priority thrust areas in all the Five-Year Plans since 1985. The open-door policy adopted by China since the mid-1980s resulted in foreign cooperation and assistance by virtue of which China could operate fishing fleets in the Exclusive Economic Zones (EEZs) of other countries, either individually or by joint ventures. The government has also set up five Special Economic Zones (SEZs), 14 open coastal port towns, and coastal development areas to attract foreign direct investment with advanced fishing technology. This was supported by the quick development of modern infrastructure facilities. Enterprises set up in SEZs with foreign investments enjoy special rights in enterprise management, low land charges, and preferential tax treatment. Imports of machinery, equipments, components and parts, and raw materials, including semi-processed materials, by enterprises set up in EEZs with foreign investment are exempted from paying customs duty. No tax is also levied on foreign remittances. As a result of all these steps taken by the government, by 1990 foreign direct investment in the SEZs had gone up to as high as 25 per cent of total direct investment in China.

As mentioned before, because of a decline in marine fish landings both in terms of quantity and species, China has put great emphasis on inland fisheries development. Fresh water production, which was 33.7 per cent of the total catches in 1983, went up to 40.45 per cent in 1996. China also accounts for more than 35 per cent of the world inland fisheries production.[3]

China has another comparative advantage in the cost of labour engaged in fisheries. It is much lower than that in her three major competitors namely, Thailand, Korea(R), and Taiwan, but is closer to that in India and Indonesia.

Despite all the steps taken by China and all the competitive advantages that she has, China is yet to obtain a price realization of her fish products to the extent of the advanced developing nations of the world as shown earlier. This is primarily due to poor on-board and offshore fish handling facilities

[3] Zhiliu, Qian, *The Development of the Chinese Fisheries and Manpower in Aquaculture*, Agricultural Press, China, 1994.

and the existence of a large number of traditional processing and preservation plants which cannot ensure high quality products to command a competitive price in the international market. However, when we compare China's price realization with other Group B developing nations and some of the developed countries, we find that China has performed better than these countries in the price realization of fish and fishery products. As a result, China is able to have a better performance in terms of the value of exports than in terms of tonnage. During 1993–5, China's average net export was 3.02 per cent of the world exports by value in US dollar terms, while in quantitative terms it was only 1.62 per cent.

China's domestic human consumption of food is considerably high. Per capita average consumption during 1993–5 was 19.1 kg per annum, increased from 4.9 kg. in 1978. The average consumption, including non-food uses, was 95.75 per cent of production including imports in quantitative terms. With the rise in income in the liberalized regime, per capita consumption is expected to rise further. This is, perhaps, one of the primary reasons why China has put development of fishery products in the top priority category. China's recent emphasis on value-added fish products for the export market also stems from this consideration. As a lesser quantity of fish is available for export the strategy has necessarily to be based on value-added speciality products. As it stands, China is unlikely to offer any significant competition, at least in the overseas traditional fish market. It would not be too surprising to find China turning out to be a net importer of fish and fishery products in the near future, particularly in the fresh, frozen form.

OTHER COUNTRIES

Peru and Chile rank second and third respectively in terms of their volume and share of the world nominal catches. Peru experienced an overall cumulative growth rate of 4.96 per cent per annum during 1975–96, which accelerated to 8.44 per cent per annum during recent years. In contrast, in Chile the high cumulative growth rate during the entire period of 1975–96 has decelerated considerably during the recent years. If mean growth rates of nominal catches of these two countries during 1975–96 are considered, we find that the growth rates in Chile and Peru are 11.52 per cent and 10.45 per cent with a standard deviation of 18.5 and 34.44 per cent, respectively. This also indicates that Chile, despite having a better mean growth rate than Peru, has suffered from deceleration in the growth rate of catches during the recent years. During 1987–96, Peru experienced negative growth rates twice, while Chile experienced it five times.

India has experienced a better cumulative growth rate during the recent years (1987–96), than the last decade and so is the case with Indonesia also. Other developing countries which have suffered from a declining cumulative growth rate during recent years include Chile, Korea(R), Thailand, Malaysia, Philippines, Korea(DPR), Vietnam, and Mexico. India's case is discussed in greater detail at a later section.

Of the 14 major developing countries only two countries, namely Malaysia and Brazil, are net importers of fishery products; all the others are net exporters. In Table 3.6, percentages of net export to domestic production and also to world export of fishery products of 14 net exporters are calculated.

TABLE 3.6

Net exporters of fish and fishery products from the developing world

(average of 1993–5)

Country	Net export as % of total production	Net export as % of world exports	Per capita consumption
Mexico	7.06	0.42	11.0
Chile	6.86	2.28	28.4
Peru	0.68	0.32	23.8
India	7.04	1.55	4.4
Indonesia	14.81	2.69	17.1
Korea(DPR)	3.59	0.30	46.0
Korea(R)	3.62	0.45	50.7
Philippines	1.23	0.13	33.8
Thailand	23.50	3.88	25.9
Vietnam	10.65	0.55	13.4
China	1.46	1.63	19.1
Taiwan	38.83	2.37	38.2
Argentina	54.25	2.54	10.1
Namibia	46.73	0.66	11.2
Total	6.54	19.77	

Note: Source of basic data and other information are as in Table 3.1.

The major net exporters of the developing world are ranked in terms of their Competitive Index in Table 3.7. Argentina has the highest competitive advantage in the fishery exports among all the developing nations, and occupies fourth position among all fish exporting countries of the world.

Argentina has also registered the highest cumulative growth rate among all fish producing nations (9.2 per cent per annum during 1975–96). The country has also consistently improved its share of the world nominal catches from 0.33 per cent in 1975 to 1.09 per cent in 1996. The reason behind Argentina's high competitive advantage is her rather low level of domestic consumption of fish products, as reflected in a comparatively low per capita consumption and high percentage of net export to production, coupled with a reasonably high percentage of net export to world export.

TABLE 3.7

Country ranking of the developing world by Competitive Index

Rank	Country	Competitive Index
1	Argentina	137.80
2	Taiwan	92.03
3	Thailand	91.18
4	Indonesia	39.84
5	Namibia	30.84
6	Chile	15.64
7	India	10.91
8	Vietnam	5.86
9	Mexico	2.97
10	China	2.38
11	Korea(R)	1.63
12	Korea(DPR)	1.08
13	Peru	0.22
14	Philippines	0.16

Note: Method of calculating Competitive Index is explained earlier.

India occupies seventh position, despite having the lowest per capita consumption and a rather high percentage of net export to production, because volume-wise her share of net export to world export is among the lowest (1.55 per cent).

Among the Asian countries, Thailand and Taiwan pose a formidable challenge. Thailand consumes fish at a moderate level, leaving enough for export. Her share of the world export is highest among the developing nations. Though Taiwan has a high per capita consumption, she is able to leave enough for export (as reflected by her ratio of net export to production) because the country has a small population—about 210 lakh.

MARKET OPPORTUNITIES IN THE DEVELOPING WORLD

Besides Brazil and Malaysia, there are five more countries in the developing world who are significant net importers of fish and fish products. In Table 3.8, their net imports as a percentage of both consumption and world import are calculated.

TABLE 3.8
*Major net importers of fish and fishery products
of the developing world*

(average of 1993–5)

Country	Net import as % of consumption	Net import as % of world import	Per capita consumption (kg)
Malaysia	6.46	0.38	54.5
Brazil	24.20	1.17	6.4
Singapore	86.54	0.41	31.8
Egypt	31.40	0.64	7.3
Nigeria	51.28	1.47	5.8
Cole devoire	62.83	0.55	12.2
Congo (DR)	30.03	0.40	6.7
Total	27.42	5.02	

Note: Basic data source and other information are as in Table 3.1.

TABLE 3.9
*Country ranking of major net-importers of fish and fishery
products of the developing world by Market Opportunity Index*

Rank	Country	Market Opportunity Index
1	Nigeria	75.38
2	Singapore	35.48
3	Cole devoire	34.56
4	Brazil	28.31
5	Egypt	20.00
6	Congo(DR)	12.00
7	Malaysia	2.45

Note: Method of calculating Market Opportunity Index has been explained earlier.

Table 3.9 ranks these countries in terms of the market opportunities provided by them in the export market of the fishery products.

Nigeria provides the greatest market opportunity among the developing nations because her net import as percentages of both domestic consumption and world import are very high. Singapore comes next (though much below Nigeria) because she is almost wholly dependent on imports for her requirement of fish.

COUNTRY RANKING

Tables 3.10 and 3.11 rank all the major net importers and net exporters of the world, irrespective of their developmental stage in terms of market opportunities and market competitiveness provided by them in the fisheries product market.

TABLE 3.10
Ranking of major countries of the world providing market opportunities in fish and fisheries products

Rank	Country	Market Opportunity Index
1	Japan	519.38
2	Italy	314.18
3	Germany	243.03
4	France	195.67
5	Spain	128.78
6	Portugal	83.50
7	Belgium	77.97
8	Nigeria	75.38
9	United States	67.87
10	Switzerland	39.68
11	Transition Economics (Group of 7 as in Table 3.3)	38.12
12	Singapore	35.48
13	Cole devoire	34.56
14	Austria	33.46
15	Israel	32.83
16	Brazil	28.31

(Contd.)

Table 3.10 (contd.)

Rank	Country	Market Opportunity Index
17	Australia	21.66
18	Egypt	20.00
19	Congo(DR)	12.00
20	Finland	10.40
21	United Kingdom	9.01
22	Greece	4.89
23	Malaysia	2.45
24	Malta	2.25

TABLE 3.11

Ranking of major countries of the world in terms of their competitive advantage

Rank	Country	Competitive Index
1	New Zealand	167.38
2	Norway	158.55
3	Former USSR area	155.95
4	Argentina	137.80
5	Taiwan	92.03
6	Thailand	91.18
7	Ireland	90.31
8	South Africa	79.80
9	The Netherlands	67.91
10	Iceland	58.03
11	Canada	41.41
12	Indonesia	39.84
13	Namibia	30.84
14	Faeroe IS	28.85
15	Chile	15.64
16	Denmark	16.07
17	India	10.91
18	Sweden	7.40
19	Vietnam	5.86
20	Mexico	2.97

(Contd.)

Table 3.11 (contd.)

Rank	Country	Competitive Index
21	China	2.38
22	Korea(R)	1.63
23	Korea(DPR)	1.08
24	Peru	0.22
25	Philippines	0.16

Besides Japan in Asia, the most market opportunities in fish and fishery products are provided by the European countries. Nigeria provides a better market opportunity than the United States.

In terms of competitive advantage in the export market of fish and fishery products, only two Asian countries, namely Taiwan and Thailand, come within the first ten ranks. India ranks only 17 among the 25 countries.

INLAND WATER FISHING

As a result of the increasing emphasis placed by several fish producing countries on the development of aquaculture due to a decline in the rate of growth of marine fish production, the share of inland water fishing in the total nominal catches of fish products has increased substantially during the last two decades, as evident from Table 3.12.

TABLE 3.12
World nominal catches of fish, crustaceans,
molluscs etc. by fishing areas

(in thousand metric tons)

Year	Marine fishing area		Inland waters		Total quantity
	Quantity	Per cent of world catches	Quantity	Per cent of world catches	
1975	58,752	89.40	6961	10.60	65,713
1976	62,157	90.00	6904	10.00	69,061
1977	61,058	89.55	7122	10.45	68,180
1978	63,098	89.88	7109	10.12	70,207
1979	63,766	89.75	7281	10.25	71,047
1980	64,516	89.38	7661	10.62	72,177

(Contd.)

Table 3.12 (contd.)

Year	Marine fishing area		Inland waters		
	Quantity	Per cent of world catches	Quantity	Per cent of world catches	Total quantity
1981	66,513	89.13	8114	10.87	74,627
1982	68,303	88.92	8515	11.08	76,818
1983	68,286	88.11	9211	11.89	77,497
1984	73,914	88.06	10,018	11.94	83,932
1985	75,714	87.65	10,664	12.35	86,378
1986	81,100	87.35	11,745	12.65	92,845
1987	81,699	86.54	12,704	13.46	94,403
1988	85,676	86.47	13,410	13.53	99,086
1989	86,436	86.17	13,875	13.83	100,311
1990	82,867	84.94	14,689	15.06	97,556
1991	82,285	84.78	14,767	15.22	97,052
1992	82,534	84.12	15,579	15.88	98,113
1993	80,618	79.34	20,990	20.66	101,608
1994	85,775	78.59	23,362	21.41	109,137
1995	85,622	76.80	25,872	23.20	111,494
1996	87,073	76.98	26,045	23.02	113,118

TABLE 3.13

Growth rates of nominal catches in marine and inland fishing areas

(in percentage)

Period	Cumulative growth rate (% per annum)		Mean growth rate (% per annum)		World total areas	
	Marine	Inland	Marine	Inland	Cumulative rate	Mean rate
1975–96	1.89	6.48	1.94	6.70	2.62	2.66
1975–84	2.58	4.13	2.53	4.18	2.76	2.79
1985–96	1.28	8.46	1.43	8.59	2.48	2.56

During 1975–96, catches from the marine fishing areas increased by 48.20 per cent, but during the same period catches from inland waters increased by 274.15 per cent. The real acceleration in catches from inland waters began in 1990 as a result of the global emphasis laid on aquaculture development during the latter half of the 1980s. Figure 3.1 illustrates the

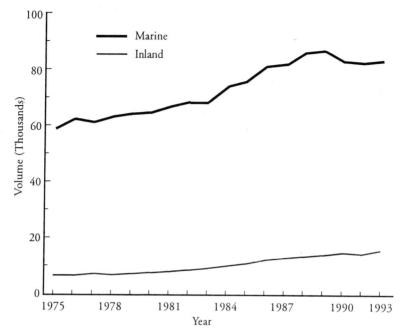

Fig. 3.1: World nominal catches of fish, molluscs, crustaceans etc. by fishing areas (in thousand metric tons)

comparative growth pattern of catches from marine and inland fishing areas of the world.

The cumulative growth rates in nominal catches of the two fishing areas for different periods along with mean growth rates are given in Table 3.13.

COMPARATIVE ANALYSIS

It may be noticed that while the cumulative growth rate of catches from the marine fishing areas is 1.89 per cent per annum for the entire period 1975–96, the growth rate increased to 2.58 per cent per annum during the mid-period of 1975–84, but has again fallen to 1.28 per cent per annum during the recent period. As if to complement this trend, exactly the opposite has happened to catches from inland waters.

The rate of deceleration in marine catches being high, the future stability in global fish and fishery production lies in faster development of aquaculture. Marine fish alone will never be able to meet the rising demand of fish products, as we have seen earlier. Inland water fisheries have to grow at a higher rate than the present trend in order to meet the supply gap of fish products.

PROJECTIONS FOR THE FUTURE

Regression analyses of nominal catch data of marine fishing areas and inland waters have given trend equations as in Table 3.14.

TABLE 3.14
Regression parameters for marine and inland water fisheries

Area	Trend equation	SE(b)	T.Value	R^2	Level of significance
Marine fishing areas	$Y_c = 58321.55 + 1441.5t$	113.82	12.66	0.89	1%
Inland waters	$Y_c = 2726.40 + 879.91t$	79.87	11.02	0.86	1%

On the basis of the above trend equations, projections of nominal catches (non-seasonalized) from the two fishing areas are given in Table 3.15.

TABLE 3.15
Projected nominal production of fish and fishery products

(in thousand metric tons)

Projected year	Projected production		
	Marine fishing areas	Inland waters	Total
2001	97,242 (78.60)	26,484 (21.40)	123,726
2002	98,682 (78.30)	27,364 (21.70)	126,046
2003	100,125 (78.00)	28,244 (22.00)	128,369
2004	101,567 (77.72)	29,124 (22.28)	130,691
2005	103,008 (77.44)	30,004 (22.56)	133,010

Note: Figures in brackets represent percentage to total.

DEMAND–SUPPLY GAP

Analysis of data on world population and per capita consumption of fish products published by the Food and Agriculture Organisation indicates that the world population is growing at a cumulative rate of 1.5 per cent per annum while per capita consumption of fish is growing at a 1.74 per cent per annum during the recent years. It has also been found that non-food usage of fish constitutes about 30 per cent of world fish production. On the basis of these three parameters and the projection of world fish production

TABLE 3.16

Projected demand–supply gap in world fish products

Year	World population (in millions)	Per capita consumption (kg./year)	Demand for human consumption [col. 2 x col. 3]	Projected total fish production [as in Table 3.15]	Non-food usage [30% of col. 5]	Available for human consumption [col. 5-6]	Demand–supply gap
(1)	(2)	(3)	(4)	(5)	(6)	(7)	(8)
2001	6130	16.08	96,963	123,726	37,118	86,608	(–) 10,355
2002	6220	16.36	101,760	126,046	37,814	88,232	(–) 13,528
2003	6315	16.65	105,145	128,369	38,511	89,858	(–) 15,287
2004	6410	16.94	108,585	130,691	39,207	91,484	(–) 17,101
2005	6505	17.23	112,081	133,010	39,903	93,107	(–) 18,974

made in Table 3.15, we have projected the demand—supply gap of fish products in Table 3.16.

Column 8 of Table 3.16 clearly indicates that with the present trend of fish production it will not be possible to meet the world demand of fishery products. There will be a general shortage of fish products in the world. The resultant effect will be as follows:

(a)   There will be a general rise in the prices of fish products;
(b)   Some countries which are hitherto marginally net exporters will turn out to be net importers of fish products;
(c)   There may be a cut-back in non-food usage of fish.

The policy focus to tackle this huge supply gap should be an organized breakthrough in fish production both in marine and inland fishing areas, with more emphasis on the latter.

ENVIRONMENTAL ISSUES

While the need of aquaculture development is well understood, we should not forget the environmental impact of its unregulated and unscientific growth which might lead to degradation of both the internal environment of aquaculture farming by disturbing the equilibrium between fish or shrimp and microbes in the water, and also the external environment by salination of land and pollution of external waterbodies through discharge of farm effluents.

Many Asian countries, in their spree to develop aquaculture, forgot the basic environmental issues and transgressed the carrying capacity of both the internal and external environments. The consequence has been widespread disease of fish and shrimp and social unrest due to water pollution. Some such countries are Republic of Korea, Taiwan, Philippines, China, and India. In some of these countries, for example India, a moratorium on aquaculture farming was also imposed for some time.

The coastal areas are most affected by uncontrolled development in the above-mentioned countries. The narrow strip of coastal belt that border continents and islands produce about 56 per cent of global aquaculture production. The high yield per unit area is the main reason behind this overgrowth. While the maximum production for inland aquaculture has been found to be below 5 tons per square kilometre on an average, the rate of production per kilometre of coastline has often reached as high as 400 tons. In their growth spree, these aquaculturists have destroyed many mangroves, which are the backbones of coastal aquaculture.

In contrast to inland aquaculture, usage of artificial feeds and drugs is very high in coastal aquaculture, which results in highly toxic organic wastes. When these effluents get discharged to coastal and other waterbodies, the pollution creates severe health hazards not only for the living organisms of water but also for the people who inhabit the coastal areas. High usage of drugs also produces more and more drug resistant bacterias and viruses in the aquasystem.

After the aquaculture crises in some Asian countries during the mid-1990s there has been a concerted move towards the concept of sustainable development in aquaculture, which aims at conservation of land, water, plant, and animal genetic resources. It is recognized that aquaculture development should be environmentally non-degrading, technically appropriate, economically viable, and socially acceptable. There is a need now to develop the safe limits of environmental parameters at the farm level and their regular monitoring for sustainable development. The Food and Agriculture Organisation also came out with a code of conduct for Responsible Fisheries in 1995. The code provides principles and guidelines for the development and management of sustainable fisheries. It contains several articles dealing, *inter alia*, with fisheries management, aquaculture development, fisheries research, fishing operations, etc. In order to maintain the equilibrium in the marine environment for sustainable growth of marine fisheries, the code has also drawn the attention of the member countries to their responsibility for the development and implementation of technology and operational management that reduces the catch of non-target species, prevents wastage through discarding, reduces the threat of endangered species, and increases the survival rate of the escaping species.

The environmental objective of aquaculture practices, as surmised from above, is that the financial gain of any such activity cannot be at the expense of the ecosystem and the society at large. This calls for suitably designing an Environmental Impact Assessment (EIA) format that would not only assess the potential impacts of an aquaculture project on the social, biological, chemical, and physical environment, but also indicate various measures to be adopted to minimize or eliminate the negative impacts. The EIA should primarily cover the impacts on (i) water course, (ii) ground water, (iii) drinking water, (iv) agriculture, (v) soil and salination, and (vi) green-belt development. The EIA should be followed by an assessment of the environmental carrying capacity of a site where an aquaculture project is being set up.

CONCLUSION

During the past two and a half decades, the terms of trade in fish products have shifted in favour of the developing countries. This has been due to the deceleration in the rate of growth of fish production in the developed nations. The process is unlikely to be reversed in the foreseeable future. It has opened up tremendous opportunities to the developing world in the expansion of their market share. The opportunity was seized by the emerging economies of the South East Asian countries, which are now the leading exporters of fish products. They were able to do it by quickly upgrading their technology in both fishing gears and fish processing and preservation. The latter has provided the physical power to carry fish stocks for better price realization.

There has been a general decline worldwide in the overall growth rate of fish production while the consumption of fish products is on the rise due to the increase in population and growth in national income. If these trends continue there will be a severe shortage of supply of fish products. Marine fish production alone cannot meet this supply gap. The need of the hour is to develop aquaculture on a fast track and to have a breakthrough in marine fishing by encouraging deep-sea fishing efforts.

While the benefit from aquaculture is well understood, the economic gain cannot be at the expense of the ecosystem and the society at large. Unscientific and uncontrolled growth in aquaculture damage the living environment of both the human and other eco-friendly species. If proper checks and balances are not maintained, social tensions may often rise to such an extent as to call for a ban or moratorium on aquaculture activities, as has happened in some of the developing nations of South East Asia and India, during the recent past.

4

Overfishing, Depletion, and Yield Potential

INTRODUCTION

Often there are some distinct changes in the pattern of catches of different species of fish that indicate overfishing of some species, leading to depletion of their stocks and ultimate extinction. This chapter first examines the characteristics of catch pattern of major groups of fish to find out whether this phenomenon is already occurring in any of the species group. Based on some of these findings an attempt is then made to calculate the yield potential of some of the important fisheries.

CATCH PATTERN OF MAJOR GROUPS OF FISH SPECIES

Eleven major groups of fish species, which constitute more than 80 per cent of the world fish trade, are considered here. There is no single yardstick by which overfishing and/or depletion of a fish species can be measured. Several parameters, such as peak catches, average growth rate, seasonal variations, and fluctuations of catches are to be considered simultaneously to arrive at a conclusion. In Table 4.1, the peak catches of major fishery groups of the world are given.

FLOUNDERS AND COD GROUPS

The peak level production of the flounders group of fish species, which includes halibats, soles, etc., has remained almost stagnant in the first two periods. The same stagnancy is also seen in the third and fourth period, though peak catches during these two periods are higher than in the first two periods. In the last period the peak catch is at the lowest. The cumulative growth rate of this group has long been negative in all the periods (– 1.34

TABLE 4.1

Peak level production of major groups of fish species in the world

(in thousand metric tons)

Period	No. of years	Flounders	Cods	Herrings, Sardines	Mackerels	Jacks	Tunas	Squids	Lobsters	Shrimp-prawns	Fresh water crustaceans	Shark
I 1973-7	5	1255	12,667	15,249	4302	8784	2391	1233	204	1672	93	595
II 1978-82	5	1257	10,982	17,867	4607	8116	2796	1639	191	1714	111	630
III 1983-7	5	1351	13,787	23,984	4267	8819	3646	2335	229	2380	247	668
IV 1988-92	5	1342	16,636	24,800	3868	10,358	4588	2738	225	2913	289	698
V 1993-6	4	1093	10,712	25,836	5137	11,135	4705	3038	227	2471	475	759

Source: Calculated from different issues of *Handbook of Fisheries Statistics'* Food and Agriculture Organisation, Rome.

per cent per annum during 1973–96, mean rate: – 1.08 per cent. The coefficient of variation (CV), which measures fluctuations, is low but negative (–13.14 per cent)[1] which, in the light of the above findings, should indicate a stagnancy in the catches.

During 1974–96, the flounders group registered negative growth rates as many as fourteen times with an average variation of –5.70 per cent per annum, while positive growth rates occurred nine times with an average variation of 6.11 per cent. The net average variation (NAV) is, therefore, 0.41 per cent.

The above parameters indicate that the flounders group of fish species is suffering from depletion phenomenon.

For the cods group, which includes hakes, haddocks, etc., the peak period catches of the fifth period is almost equal to that of the second period. The peak catches of these two periods are also the lowest among all the periods. The peak catches in the fourth period is higher than that of all other periods. Closely following the peak level catches of all the five periods, it can be noticed that there is an element of seasonality in the catch behaviour of this species. Bumper catches occur every alternate year. As for the flounders group, the cumulative growth rate for cods group has also remained negative during 1973–96, though at a low of –0.48 per cent per annum (mean rate: –0.09 per cent). The CV is also low, though negative (–10.60 per cent).

During 1974–96, negative growth rates for the cods group occurred only nine times as compared to fourteen times for the flounders group with a NAV of –3.07 per cent. When all the parameters are considered together, particularly the seasonality aspect, it can be concluded that the depletion phenomenon has not yet occurred in the fish species of the cods group.

HERRINGS VS. MACKERELS GROUP

It is usually said that the fish species of herrings group (which include sardines etc.) and mackerels group (which include anchovies like, snoeks, cutlass, etc.) are subjected to alternate seasonality between their catches. The same behaviour is observed in the analysis here. An examination of the peak catches of these two groups for the five periods in Table 4.1 shows that the periods in which herrings is at its top peak production are those in which mackerels production is at its lowest peak.

For herrings, the peak catches of every period exceed that of the former

[1] Although a high CV denotes high fluctuations which renders production and market planning difficult, it also signifies that the fish stock is not subject to the depletion phenomenon. A very low CV, on the other hand, denotes that a particular fish species is near full exploitation or reaching a saturation point.

period. This is not true of mackerels group for periods three and four, but the peak catches of the fifth period has exceeded all other peak period catches.

The cumulative growth rate of catches of herrings group has been 2.92 per cent per annum during the entire period 1973–96. However, in the more recent period 1983–96 it has come down to 1.90 per cent though in the earlier period 1973–82 it was 5 per cent. In the case of mackerels group, the rate of growth per annum for the entire period has been 1 per cent, which has increased to 2.61 per cent during the recent period, though, in the early period it was at a negative figure, –0.66 per cent. The growth pattern of catches of these two groups of fishes are diametrically opposite to each other, which also supports the earlier hypothesis.

The fluctuation in catches for herrings group (CV equal to 10.59 per cent) is lower than that for the mackerels group (CV equal to 21.70 per cent). Negative growth rates have occurred nine times in the former group while for the latter group this has happened twelve times during 1974–96. However, the averages of negative and positive growth rates are about the same for both the groups. All these suggest that neither of these two groups of fish species suffers from any depletion phenomenon.

JACKS, TUNAS, AND SQUIDS GROUPS

Table 4.1 shows that the peak catches of the jacks, tunas, and squids groups are on the rise in succeeding periods, except in the second period for the jacks group but here too the peak catches of the next period are more than that of the immediately preceding period. A detailed examination of the catch behaviour of tunas and squids reveals that their peak period catches occur in every last year of all the five periods. This finding indicates the existence of seasonality in the catches of these two species.

These three groups of fish species enjoy the highest cumulative growth rates among all the groups (2.98 per cent, 3.44 per cent and 4.65 per cent per annum, respectively), except freshwater crustaceans (7.71 per cent). The degree of fluctuations (measured by the coefficient of variation) in the catches of these three groups are 18.02 per cent, 28.37 per cent, and 33.56 per cent, respectively, which are also amongst the highest of all the groups. The high degree of fluctuations in the latter two groups, namely tunas and squids, makes them risky for production and market planning, but at the same time this indicates the likely nature of these two species. Negative growth rate occur in these three species nine, four and five times, respectively, while the net average variations (NAVs) are positive for all of them, with the jacks group leading (3.62 per cent), followed by squids group (2.51 per cent) and tunas group (1.28 per cent).

The above findings clearly indicate that there is no evidence of depletion of stock in any of these three groups of fish species.

LOBSTERS, SHRIMP–PRAWNS AND FRESH WATER CRUSTACEANS

A detailed scrutiny of the peak-period catches of the lobsters, shrimp–prawns, and freshwater crustaceans groups of fish species over 24 years (1973–96) suggests that production of lobsters group remains stagnant for all the five periods. For the shrimp–prawns group and freshwater crustaceans, peak catches are found to be occurring in the last year of every five periods considered in Table 4.1 except for the first period of freshwater crustaceans when the peak catch has occurred in its third year. This finding is similar to that of tunas and squids groups which suggests seasonality in their catches.

The cumulative growth rates of all these three groups of fish species are positive at 1.35 per cent, 2.87 per cent and 7.7 per cent per annum, respectively, during 1973–96. The growth parameter of freshwater crustaceans is highest among all the fishery groups studied here. During the recent years (1983–96), the rate of growth has been 11.76 per cent per annum. This exemplary growth rate is due to the emphasis laid on aquaculture throughout the world since the mid-1980s.

Volatility in catches, as measured by the coefficient of variation (CV), for lobsters, shrimp–prawns, and freshwater crustaceans is 11.31 per cent, 24.19 per cent, and 57.61 per cent, respectively, for the entire period, 1973–96. High volatility for freshwater crustaceans is due to the 'quantum' jump in aquaculture production since 1980. During the last ten years (1987–96), we find a more stable situation with a CV of 26.25 per cent. Negative growth rates occurred ten times in the lobster group, followed by the freshwater crustaceans group. It is least in the shrimp–prawn group (five times). The NAV is a small negative figure in the shrimp–prawn group (–1.59 per cent) while it is a high positive figure in the freshwater crustaceans group (8.73 per cent). For the lobsters group, the NAV is a small positive figure (1.73 per cent).

An evaluation of all the parameters discussed above indicates that except for the lobsters group, which is reaching its plateau of catches, there is no evidence to suggest a depletion of stocks in the other two groups.

SHARK GROUP

A study of the peak level of catches of this group of species for the five periods reported in Table 4.1 indicates that it is reaching a new higher peak in every subsequent period. Detailed scrutiny of year to year catches for

1973–96 has revealed that except in the first period (where peak catches occurred twice), all the peak-level catches occurred in the last year of every period. This suggests the presence of seasonality in the catch behaviour of the shark group of fish species.

The cumulative growth rate of production of the shark group has been 1.06 per cent per annum for the entire period (1973–96); it has fallen to 0.64 per cent per annum during the mid-period, but once again has risen to a high of 2.02 per cent during the recent period (1983–96). Volatility in catches is low at 10.33 per cent. Negative growth rates occurred only 6 times during 23 years, while the NAV is a small negative figure (–1.23 per cent).

All the above findings suggest that the sharks group indicates no sign of depletion of stocks.

YIELD POTENTIAL

There are several methods of calculating the maximum sustainable yield (MSY) potential of a fish or a group of fish species, but the one that is close to fishermen's 'gut-feeling' is based on the principles of bio-economics. As the sea belongs to the 'commons' with no ownership attributed to any individual (except for the 200-mile Exclusive Economic Zone belonging to a coastal state), the sea ceases to be considered as an asset, like air, which needs to be preserved or conserved. This is the fundamental difference between marine fishing and agriculture, or for that matter mining. It leads the fishermen to implicitly apply an infinite discounting rate to the sea asset. The level of exploitation of the sea-resources, therefore, depends only upon the finite discount rate on the capital invested in boats and other fishing equipment. The higher this discount rate is, the higher will be the level of exploitation. The check-point, however, is provided by the marginal revenue concept of economics. If the marginal cost of an additional catch is more than the marginal revenue, then such an additional catch will not be sustainable for the fishermen. The marginal cost of fishing means principally the cost of an additional unit of fishing effort. The additional fishing effort will not be sustainable if the price realization from such an effort is lower than its cost, despite the size of the catch. Or the size of catch is such that additional expected revenue would not be forthcoming despite no alteration in unit price of the catch. The former emanates from the principles of market economics, based on demand and supply, and the latter is based on the biological principles of breeding and mortality rate of different fish species and their migratory habits.

TWO TYPES OF YIELD

There will, therefore, be two types of yield: economically sustainable yield and biologically sustainable yield. The former level may be reached even when there is an abundance of fish, and more often when the situation is such. A fall in the price of a fish, may cause the marginal revenue to fall below the marginal cost of a fishing effort and, hence, fishermen will stop fishing despite the abundance of stock.[2] We call this situation underfishing. On the other hand, so long as the marginal revenue remains higher than the marginal cost of the fishing effort, fishermen will continue to fish without regard to the depletion of fish stock—more so in such a situation, because with a fall in catch, the price of the fish will rise, demand remaining constant. We call this situation overfishing. A situation will, however, be reached soon when the demand will fall due to rising prices. Consequently, the marginal revenue will start declining and reach the equalizing level of the marginal cost, when further fishing effort will stop. But the danger of allowing this free play of economic forces in fishing is that, in the meantime, there is a chance that a particular fish species might become extinct due to overfishing. This calls for the regulation of fishing efforts. The maximum sustainable yield will, therefore, be a trade-off between the economic forces controlled by the market and biological forces that determine the fish population.

AN ALTERNATIVE APPROACH

We have seen that year to year fluctuations in catches are very high in all the fish types. This is a general phenomenon of the fishing activity. On scrutinizing the catch data of different groups of fish species or individual fish types, it is found that a kind of seasonality operates in fish catches, for example, a particular fish will reach its maximum catch every fifth year. The catch may fall thereafter till it reaches the next fifth year. Examples of such fishes, as found from our study, are discussed earlier. Besides this, two other types of peak-catch behaviours are observed. In some types or groups of fishes, the peak catch of every period is found to exceed that of the immediately preceding period. In the case of some other fishes, exactly the opposite behaviour is observed. There is also a third group of fishes who do not conform to either of these exact behavioural patterns; their peak catches also fluctuate. These peak-catch behaviours obviously reflect the interplay of both economic and biological forces, indicating the maximum sustainable

---

[2] However, it is found that fishermen continue to go on fishing despite a fall in the marginal revenue. This is due to the implicit cost of keeping the vessel and men idle.

yield potential for any fish type or group. This indication can be captured by discerning the trend factor of peak catches over several periods. For this purpose we have followed the principles of moving average which lays more emphasis on recent period occurrences. The trend factor is then multiplied with the peak catch figure of the latest period to determine the MSY potential of a fish or group of fish species. The model adopted for this purpose is as follows:

$$MSY_p = Q_n \left\{ \left(\frac{Q_1 + 2Q_2 + 3Q_3 \ldots\ldots + nQ_n}{Q_1 + Q_2 + Q_3 \ldots\ldots + Q_n} \right) \frac{2}{n+1} \right\}$$

where,

Q	=	Peak catches of a period;
Q_n	=	Peak catches of the latest period;
n	=	Number of periods.

With the help of the above model the maximum sustainable yield potential of different groups of fish species are calculated in Table 4.2.

TABLE 4.2

Maximum sustainable yield potential of
major groups of fish species

(in thousand metric tonnes)

Species group	Potential
Flounders, halibats, soles etc.	1100
Cods, hakes, haddock etc.	11,000
Herrings, sardines, anchovies etc.	28,000
Mackerels, snoeks, cutlass etc.	5200
Jacks, mullets etc.	11,700
Tunas, bonitos, billfish etc.	5100
Squids, cuttlefish, octopus etc.	3500
Lobsters–spiny, rock, squat etc.	230
Shrimps, prawns	2700
Fresh water crustaceans	600
Sharks, rays, chimaeras etc.	800

TABLE 4.3

Summary of findings in respect of the major fish of the world

Name of fish	Cumulative growth rates (% p.a.)			Mean growth rate (%)	Fluctuations in catches				
	1973–96	1973–82	1983–96		Variation from mean (CV%)	Negative seasonal occurrences (no.)	Average seasonal variation (NASV %)	Succeeding peak-period catch exceeding the preceding period (no.)	Existence of depletion phenomenon
Alaska Pollock	(–) 0.49	2.01	(–) 3.92	(–) 0.11	(–) 8189.25	10	(–) 0.86	2**	Observed
Japanese Pilchard	(–) 0.98	24.45	(–) 22.03	3.79	864.94	9	0.63	2**	Observed
Chilean Pilchard	9.36	38.25	(–) 10.10	17.83	300.12	11	38.35	2**	Observed
Capelin	(–) 1.82	(–) 0.14	0.82	3.89	937.30	10	5.17	0**	Not observed
Chilean Jack Mackerel	13.63	21.80	8.37	16.59	171.23	6	16.54	3*	Not observed
Chub Mackerel	0.49	(–) 1.15	0.76	2.75	810.60	11	7.29	0*	Not observed
Atlantic Cod	(–) 2.84	(–) 2.16	(–) 4.19	(–) 2.62	(–) 251.53	14	(–) 1.31	0**	Observed
Atlantic Herring	2.03	(–) 0.02	4.22	2.76	439.82	8	0.89	2*	Not observed
Peruvian Anchovy	4.79	(–) 1.42	6.01	69.94	313.82	7	79.19	2**	Not observed
Haddock	(–) 1.81	(–) 3.35	(–) 11.46	(–) 0.73	(–) 1992.79	10	(–) 2.68	1*	Observed
Skipjack Tuna	4.85	5.24	3.20	5.55	218.48	8	6.16	3*	Not observed

(Contd.)

Table 4.3 (contd.)

Name of Fish	Cumulative growth rates (% p.a.)			Mean growth rate (%)	Fluctuations in catches				Existence of depletion phenomenon
	1973–96	1973–82	1983–96		Variation from mean (CV%)	Negative seasonal occurrences (no.)	Average seasonal variation (NASV %)	Succeeding peak-period catch exceeding the preceding period (no.)	
European Pilchard	(–) 0.37	(–) 1.44	0.08	0.67	2128.30	10	0.20	1	Observed
Yellowfin Tuna	3.19	3.58	2.19	3.45	212.94	8	3.74	3 *	Not observed
Indian Oil Sardine	(–) 2.88	(–) 2.72	(–) 3.41	(–) 2.03	(–) 628.15	12	(–) 1.09	2 **	Observed
Indian Mackerel	7.11	6.31	9.90	9.65	245.96	9	13.82	2 *	Not observed
Blue Whiting	11.42	26.10	(–) 2.54	18.29	241.83	9	28.52	1 *	Reaching plateau

* Peak-catches in the last two-year period exceeded that of the previous period.

** Peak-catches in the last two-year period declined from that of the previous period.

Source: Calculated from catch data reported in various issues of *Handbook of Fisheries*, Food and Agriculture Organisation, Rome.

PRODUCTION PATTERN OF IMPORTANT FISH

In line with the foregoing analysis we have also investigated the production characteristics of sixteen fish that individually dominate the world market for fish. The results of our investigation are reported in Table 4.3.

YIELD POTENTIAL FOR MAJOR FISH TYPES

On the basis of peak-catch performances of various fish types we have calculated their maximum sustainable yield potentials, as shown in Table 4.4. The methodologies for calculation of MSY are the same as discussed earlier. The last period consisting of only two years is ignored for the purpose of this calculation. The derived trend factor is multiplied with the fourth period peak catches to arrive at the MSY. In the case of the Japanese pilchard, the unusual first period peak-catch has been ignored.

TABLE 4.4

Maximum sustainable yield (MSY) potentials for major fish types of the world

(in thousand metric tons)

Fish type	Potential yield (MSY)
Alaska Pollack	5800
Japanese Pilchard	4600
Chilean Pilchard	4400
Capelin	1900
Chilean Jack Mackerel	5000
Chub Mackerel	1300
Atlantic Cod	1400
Atlantic Herring	2000
Peruvian Anchovy	15,500
Haddock	300
Skipjack Tuna	1700
European Pilchard	1600
Yellowfin Tuna	1350
Indian Oil Sardine	330
Indian Mackerel	400
Blue Whiting	525

CONCLUSION

The study has revealed that the major fishery resources belonging to the temperate regions of the world are suffering from a depletion phenomenon. This region is populated by the developed countries of the world. On the contrary, most of the fish species of the coastal countries of the developing world are generally not subject to such a phenomenon. This calls for an alternative long-term marketing strategy for the developing countries to popularize their types of fish in the countries of the developed world.

At the same time, conservation of fish species is the order of the day. Appropriate regulations and joint efforts by both developed and developing nations are necessary to avert depletion due to overfishing. Developing countries should be particularly careful and draw lessons from what happened to the species belonging to the temperate regions of the developed countries, as otherwise the same phenomenon may be repeated in the developing countries also owing to the pressure of market forces.

5

Disposition of World Fish Production and Import Trade

INTRODUCTION

The consumption of fish has a high cultural orientation. Despite the advancement of technology relating to fish processing and preservation, the majority of fish-eating people continue to demand fish in the fresh form. There is no sign of abatement of this demand; rather, it is increasing with the passage of time. The second best alternative product is the frozen fish, which attempts to maintain the freshness of fish as much as possible. Despite the hopes generated for a wide expansion of the processed fish market with the advent of modern food (fish) processing technology, the demand for processed fish has not expanded to expectations; rather it has shrunk during recent times. Table 5.1 shows the disposition of fish catches supporting the above observations.

The share of fresh fish in the total disposition of fish has increased from 19.5 per cent in 1983 to 32.3 per cent in 1996, while the rest of the fish forms have barely been able to maintain their share. It may be pointed out that while the world fish production expanded by 56.15 per cent between 1983 and 1996, the disposition of fish in fresh form expanded by a whopping 158.27 per cent. The expansion of different forms of fish production between 1983 and 1996 is shown in Table 5.2.

The indication is clear. The dominant mode of consumption of fish shall continue to be in fresh form, followed by frozen or chilled fish—the nearest form of fresh fish. Fish in fresh and frozen or chilled forms together have a share of more than 55 per cent in the total disposition of fish. When non-food usages of fish are excluded, the share increases to 73.5 per cent of all human consumption of fish.

TABLE 5.1
Disposition of world fish production

(in thousand metric tons)

Year	Fresh	Frozen	Cured	Canned	Reduction for non-human consumption	Miscellaneous purposes
1983	15,114 (19.5)	19,099 (24.6)	9846 (12.7)	10,970 (14.2)	21,018 (27.1)	1450 (1.9)
1984	16,089 (19.2)	20,481 (24.4)	9878 (11.8)	11,382 (13.5)	24,552 (29.3)	1550 (1.8)
1985	17,141 (19.8)	20,846 (24.1)	10,265 (11.9)	11,441 (13.3)	25,385 (29.4)	1330 (1.5)
1986	19,544 (21.1)	22,227 (23.9)	10,315 (11.1)	11,527 (12.4)	27,432 (29.6)	1800 (1.9)
1987	23,018 (24.1)	22,646 (23.7)	10,203 (10.7)	11,766 (12.3)	26,420 (27.6)	1550 (1.6)
1988	24,751 (24.7)	23,711 (23.6)	10,247 (10.2)	12,146 (12.1)	27,862 (27.8)	1650 (1.6)
1989	24,150 (23.8)	24,047 (23.7)	10,564 (10.4)	12,651 (12.4)	28,653 (28.2)	1600 (1.6)
1990	22,757 (23.0)	24,580 (24.8)	10,859 (11.0)	12,921 (13.1)	26,392 (26.7)	1500 (1.5)
1991	21,751 (22.0)	24,325 (24.6)	10,961 (11.1)	13,144 (13.3)	27,074 (27.4)	1600 (1.6)
1992	26,164 (25.7)	24,687 (24.3)	10,163 (10.0)	12,505 (12.3)	26,409 (26.0)	1800 (1.8)
1993	26,641 (25.3)	26,138 (24.8)	10,366 (9.9)	12,826 (12.2)	27,435 (26.1)	1800 (1.7)
1994	29,319 (25.8)	26,752 (23.6)	11,170 (9.8)	12,751 (11.2)	31,666 (27.9)	1800 (1.6)
1995	34,648 (29.5)	27,286 (23.3)	117,61 (10.0)	12,799 (10.9)	28,984 (24.7)	1800 (1.5)
1996	39,035 (32.3)	27,532 (22.8)	116,21 (9.6)	12,428 (10.3)	28,494 (23.5)	1900 (1.6)

Note: Figures in brackets represent percentage share of total disposition of fish.
Source: Fishery Statistics Year Book, Food and Agriculture Organisation, Rome, various issues.

TABLE 5.2

*Expansion in the disposition of fish production
between 1983 and 1996*

(in per cent)

World production	56.15
Fresh	158.27
Frozen	44.15
Cured	18.03
Canned	13.29
Reduction	35.57
Miscellaneous purposes	31.03

WORLD TRADE OF FISH IN DIFFERENT FORMS

This particular direction in the consumption pattern of fish will obviously be reflected in the world trade in fish. In this chapter, the Food and Agriculture organisation's (FAO's) classification and grouping of fish products into four groups is used. These are (i) 'Fish—Fresh, Chilled or Frozen', which include live fish, fish fillets, and Individually Quick Frozen (IQF) materials; (ii) 'Shellfish (Crustaceans)—Fresh, Chilled or Frozen', which include crabs and all peeled shrimps, prawns and lobsters in different forms like peeled undeveined (PUD), peeled deveined (PD), peeled and cooked (PC) (tail on or without), tray packs, all IQF and Block Frozen materials; (iii) 'Fish—Salted, Dried or Smoked', which include fish fillets salted or in brine and (iv) 'Fish—Prepared, Preserved, and Others' which include cooked fish salad, squids salads, fish cutlets, prawn cutlets, stuffed crabs, fish burger, fish pickles, breaded shrimps, surimi, sushi seafood mix, and canned items. Table 5.3 and Fig. 5.1 show world imports of fish in these four major forms.

Table 5.3 shows that World imports of both fish and crustaceans (shellfish) in fresh, frozen or chilled form dominate the market share, together constituting nearly 77 per cent of world imports in 1996. Imports of fish in prepared and preserved form has barely maintained its market share while fish in salted, dried, or smoked form has lost its share of the market. In terms of both cumulative growth rate and mean growth rate, fish in fresh, chilled or frozen form has fared much better than all the other fish forms.

While the market domination of fish in fresh, frozen, or chilled form is clearly evident from all the parameters, one should not ignore the fact that though fish in prepared and preserved form has not been able to increase its

TABLE 5.3
World import of fish in different forms

(in US million dollars)

Year	Fish: fresh chilled or frozen	Shellfish: fish, chilled or frozen	Fish: salted, dried or smoked	Fish: prepared, preserved and others	Total
1979	5432 (39.14)	4743 (34.18)	1275 (9.19)	2428 (17.50)	13,878
1980	5579 (39.24)	4497 (31.63)	1240 (8.72)	2900 (20.40)	14,216
1981	6061 (40.22)	4690 (31.12)	1340 (8.89)	2979 (19.77)	15,070
1982	6113 (40.46)	5156 (34.13)	1152 (7.63)	2686 (17.78)	15,107
1983	5695 (37.50)	5650 (37.20)	1111 (7.31)	2732 (17.99)	15,188
1984	6017 (38.98)	5704 (36.95)	1006 (6.52)	2711 (17.56)	15,438
1985	6843 (40.48)	6010 (35.55)	1145 (6.77)	2908 (17.20)	16,906
1986	9254 (41.05)	8106 (35.96)	1405 (6.23)	3777 (16.76)	22,542
1987	11,942 (41.92)	10,112 (35.49)	1896 (6.65)	4540 (15.94)	28,490
1988	14,459 (43.18)	11,444 (34.18)	2080 (6.21)	5499 (16.42)	33,482
1989	14,605 (43.46)	11,423 (33.99)	1993 (5.93)	5586 (16.62)	33,607
1990	16,937 (45.21)	12,095 (32.29)	2386 (6.37)	6044 (16.13)	37,462

(Contd.)

Table 5.3 (contd.)

Year	Fish: fresh chilled or frozen	Shellfish: fish, chilled or frozen	Fish: salted, dried or smoked	Fish: prepared, preserved and others	Total
1991	19,038 (45.78)	13,226 (31.82)	2515 (6.05)	6796 (16.35)	41,565
1992	19,416 (45.53)	13,558 (31.79)	2548(5.97)	7123 (16.70)	42,645
1993	18,463 (44.08)	14,122 (33.72)	2251(5.37)	7050 (16.83)	41,886
1994	20,503 (42.90)	16,839 (35.23)	2508(5.25)	7946 (16.62)	47,796
1995	22,716 (42.83)	18,254 (34.42)	2912(5.49)	9152 (17.26)	53,034
1996	23,522 (44.00)	17,399 (32.54)	2913 (5.45)	9629 (18.01)	53,463
(a) Cumulative growth rate (% p.a.)	7.91	7.96	1.40	6.52	7.18
(b) Mean growth rate(%)	9.53	8.37	5.74	8.87	8.65
(c) Fluctuations (CV %)					
(a) 1979–96	50.02	45.43	33.59	45.31	46.51
(b) 1987–96	19.39	19.13	13.92	22.18	19.19

Notes: (i) Figures in brackets represent per cent of world import;
(ii) Computed from various issues of International Trade Statistics Year Book Volume II—Trade by Commodity. World import data presented here differ from that published by FAO.

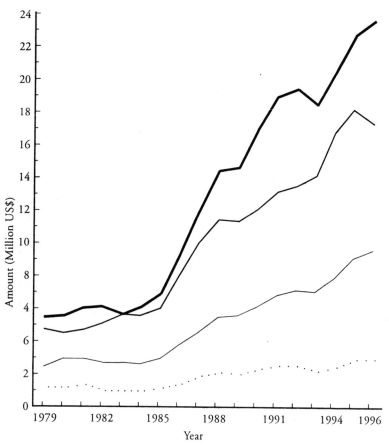

*Fig. 5.1: World import of fish products in different forms
(million US dollars)*

market share (rather, the share has shrinked from the year 1980), in absolute terms the size of the market has increased from US$2428 million in 1979 to US$9629 million in 1996. While the world import market has expanded by about 285 per cent between 1979 and 1996, the market for fish in prepared and preserved form has expanded by about 296.50 per cent during the same period, which is a commendable performance. This indicates that the demand for this type of fish product is expanding at a fast rate (though it is much below that of fish in fresh, chilled or frozen form, which expanded by more than 333 per cent during the same period). In terms of market expansion, it has also beaten crustaceans in fresh, chilled or frozen form, which grew by about 267 per cent—much below the rate of expansion of the world market during the same period. It is likely that the prepared and preserved form is making inroads into the fresh, chilled or frozen market of crustaceans because of its amenability to processing and consequent high value addition.

PRICE REALIZATION

Table 5.4 examines the price realization of different forms of fish products.

TABLE 5.4
Unit price realization of different forms of fish products

(per metric ton by US $1000)

Year	Fish: fresh, chilled or frozen	Crustaceans and molluscs: fresh chilled or frozen	Fish: dried salted or smoked	Fish: canned	Crustaceans and molluscs: canned
1993	2.10	5.06	4.61	3.45	6.68
1994	2.17	5.90	4.52	3.46	6.85
1995	2.28	6.43	4.91	3.70	7.60
1996	2.30	5.89	4.59	3.63	7.26
Average	2.22	5.82	4.66	3.57	7.12

Source: Fisheries Statistics: Commodities Year Book, Food and Agriculture Organisation, Rome, various issues.

Although price realization figures, as given in Table 5.4, are strictly not comparable among different forms of fish products because these are calculated on weights of the individual products (1 metric ton of fresh fish cannot be strictly compared with 1 metric ton of processed/canned fish), the fact of high value addition in prepared or preserved forms of fish and crustaceans etc. cannot be ignored. Value addition is much higher in canned fish (about 60 per cent) as compared to canned crustaceans and molluscs (about 22 per cent).

Value addition, however, costs money in terms of capital investment in sophisticated processing plants with high level quality control mechanisms and marketing infrastructure. While developing nations cannot ignore the processed fish markets which provide several niche segments of speciality fish items, a capital starved developing economy, like India, which lacks managerial and technical expertise, should move cautiously in these niche markets which are located mainly in the countries of the developed world. The optimal marketing strategy should be to strengthen the marketing of fish in fresh or frozen form, keeping the speciality fish market reserved for high net worth companies with the necessary managerial and technical expertise. Joint ventures with international companies can also be explored in this market segment.

The following sections analyses the status of importing continents/ countries in different forms of fish products.

IMPORT OF FISH IN FRESH, FROZEN OR CHILLED FORM

CONTINENT ANALYSIS

Table 5.5 presents the import data of fish in fresh, frozen or chilled form by five continents.

America and Europe

Table 5.5 shows that the share of America (dominated by the United States) in the world import of fish in fresh, frozen or chilled form has gone down substantially (almost halved) during the period under consideration, 1979–96. Since 1988 the fall has been steep, which is also reflected by the cumulative growth rate falling to a level of 1.30 per cent per annum during 1988–96 from a level of 8.43 per cent per annum during 1979–88. This indicates a clear switch in preference by the fish eating people of America. This is also supported by the fact that the share of this particular fish product among all the fish products imported by America went down from 42.06 per cent in 1979 to 35.74 per cent in 1996 while the share of the prepared and preserved form of fish product increased from 13.71 per cent to 17.20 per cent during the same period.

TABLE 5.5
Import of fish in fresh, frozen or chilled form by continent

(in million US dollars)

Year	America	Europe	Asia	Africa	Oceania
1979	1418 (26.10)	2458 (45.25)	1327 (24.43)	147 (2.71)	83 (1.53)
1980	1373 (24.61)	2764 (49.54)	1143 (20.49)	177 (3.17)	122 (2.19)
1981	1552 (25.61)	2506 (41.35)	1588 (26.20)	301 (4.97)	114 (1.88)
1982	1464 (23.95)	2547 (41.67)	1743 (28.51)	259 (4.24)	100 (1.64)
1983	1462 (25.67)	2207 (38.75)	1715 (30.11)	203 (3.56)	108 (1.90)
1984	1519 (25.25)	2144 (35.63)	1984 (32.97)	263 (4.37)	107 (1.78)
1985	1730 (25.28)	2447 (35.76)	2287 (33.42)	265 (3.87)	114 (1.67)
1986	2133 (23.05)	3467 (37.46)	3291 (35.56)	255 (2.76)	107 (1.16)
1987	2740 (22.94)	4774 (39.98)	4018 (33.65)	272 (2.28)	136 (1.14)
1988	2938 (20.32)	5335 (36.90)	5846 (40.43)	273 (1.89)	316 (2.19)
1989	2673 (18.30)	5375 (36.80)	5715 (39.13)	317 (2.17)	246 (1.68)
1990	2549 (15.05)	7456 (44.02)	6103 (36.03)	361 (2.13)	231 (1.36)
1991	2653 (13.94)	8213 (43.16)	7222 (37.95)	432 (2.27)	205 (1.08)
1992	2477 (12.76)	8244 (42.46)	8007 (41.24)	355 (1.83)	153 (0.79)
1993	2669 (14.46)	6855 (37.13)	8287 (44.88)	323 (1.75)	150 (0.81)
1994	2882 (14.06)	7447 (36.32)	9260 (45.16)	367 (1.79)	171 (0.83)
1995	3198 (14.08)	8379 (36.89)	9993 (43.99)	415 (1.83)	183 (0.80)
1996	3258 (13.85)	8672 (36.87)	10,047 (42.71)	598 (2.54)	183 (0.78)

Cumulative growth rate (% p.a.)

(a) 1979–96	5.02	7.70	12.65	8.6	4.76
(b) 1979–88	8.43	8.99	17.91	7.12	16.01
(c) 1988–96	1.30	6.26	7.00	10.30	(–) 6.60

Note: Figures in brackets represent percentage to total world import of this form of fish product.

Source: As in Table 5.3.

The same switch in preference has occurred in Europe also, but is not as large as in America. The cumulative growth rate during 1988–96 fell as compared to the period 1979–88, but the fall was not as high as in America. The share of this fish product amongst all the fish imports of Europe went down from 48.84 per cent in 1979 to 45.72 per cent in 1996. While the world market for this product increased by 333 per cent between 1979 and 1996, for America the expansion is only about 130 per cent, while for Europe it is about 253 per cent. America and Europe, which together imported more than 71 per cent of the world import of this form of fish product in 1979, now account for about 50 per cent of this world import.

Asia

The loss in American and European markets has been more than offset by a substantial expansion of this form of fish product in the Asian market. Its share was about 24 per cent in 1979, which increased to about 43 per cent in 1996. The total market has also expanded by 657 per cent between 1979 and 1996, nearly double the rate of expansion of the world market for this product, though the cumulative growth rate has fallen during the recent period (1988–96). This product now constitutes 44.39 per cent of the total imports of fish products in Asia, which was only 28.37 per cent in 1979.

Africa and Oceania

Africa, though a comparatively small importing continent, has been able to hold its market share by maintaining a high growth rate. Between 1979 and 1996 there has been a 307 per cent market expansion. There is also a phenomenal growth in the share of this fish product among all the fish products imported in Africa—the share, which was only 28.32 per cent in 1979, has gone up to 74.10 per cent in 1996. At the same time the share of prepared and preserved fish product has gone down from 47.21 per cent in 1979 to only 16.60 per cent in 1996.

Oceania's market of this fish product is shrinking, with a negative compound growth rate during the recent period. The share of this fish product in the total import of fish products in Oceania has also gone down from 33.07 per cent to 30.45 per cent in 1996.

COUNTRY ANALYSIS

Table 5.6 analyses the import data of the major importing countries from America and Europe.

TABLE 5.6

Import of fish in fresh, frozen or chilled form by major importers from America and Europe

(in million US dollars)

Year	United States	France	Spain	Germany	United Kingdom	Italy	Denmark	The Netherlands	Canada
1979	1267(23.32)	427(7.86)	174(3.20)	408(7.51)	382(7.03)	338(6.22)	151(2.78)	139(2.56)	67(1.23)
1980	1222(21.90)	469(8.41)	207(3.71)	476(8.53)	403(7.22)	380(6.81)	175(3.14)	147(2.63)	89(1.60)
1981	1403(23.15)	430(7.10)	220(3.63)	379(6.25)	350(5.77)	359(5.92)	153(2.52)	114(1.88)	90(1.48)
1982	1322(21.63)	435(7.12)	258(4.22)	364(5.95)	340(5.56)	366(5.99)	161(2.63)	103(1.68)	79(1.29)
1983	1342(23.56)	436(7.66)	159(2.79)	360(6.32)	311(5.46)	336(5.90)	150(2.63)	84(1.47)	86(1.51)
1984	1383(22.98)	429(7.13)	139(2.31)	339(5.63)	294(4.89)	342(5.68)	149(2.48)	87(1.45)	98(1.63)
1985	1583(23.13)	451(6.59)	175(2.56)	365(5.33)	345(5.04)	448(6.55)	172(2.51)	104(1.52)	111(1.62)
1986	1905(20.58)	647(6.99)	279(3.01)	541(5.85)	474(5.12)	541(5.85)	283(3.06)	158(1.71)	140(1.51)
1987	2461(20.61)	892(7.47)	523(4.38)	641(5.37)	571(4.78)	784(6.57)	399(3.44)	224(1.88)	167(1.40)
1988	2153(14.89)	981(6.78)	744(5.15)	707(4.89)	611(4.23)	811(5.61)	424(2.93)	247(1.71)	160(1.11)
1989	2323(15.91)	954(6.53)	761(5.21)	728(4.98)	602(4.12)	838(5.74)	399(2.73)	269(1.84)	187(1.28)
1990	2197(12.97)	1335(7.88)	1076(6.35)	1024(6.05)	851(5.02)	1110(6.55)	591(3.49)	370(2.18)	183(1.08)

(Contd.)

Table 5.6 (contd.)

Year	United States	France	Spain	Germany	United Kingdom	Italy	Denmark	The Netherlands	Canada
1991	2290(12.03)	1392(7.32)	1255(6.60)	1158(6.09)	815(4.28)	1167(6.13)	664(3.49)	432(2.27)	207(1.09)
1992	2112(10.88)	1412(7.27)	1294(6.66)	1174(6.05)	1131(5.83)	1131(5.83)	669(3.45)	453(2.33)	209(1.08)
1993	2176(11.79)	1155(6.26)	1079(5.84)	1058(5.73)	690(3.74)	899(4.87)	509(2.76)	365(1.78)	271(1.47)
1994	2251(10.98)	1208(5.89)	1176(5.74)	1278(6.23)	667(3.25)	884(4.31)	550(2.68)	426(2.08)	305(1.49)
1995	2458(10.82)	1348(5.93)	1321(5.82)	1375(6.05)	716(3.15)	941(4.14)	620(2.73)	499(2.20)	370(1.63)
1996	2511(10.68)	1416(6.02)	1383(5.88)	1366(5.81)	808(3.44)	1025(4.36)	608(2.58)	497(2.11)	384(1.63)

Cumulative growth rate (% p.a.)

	United States	France	Spain	Germany	United Kingdom	Italy	Denmark	The Netherlands	Canada
(a) 1979–96	4.11	7.31	12.92	7.37	4.51	6.74	8.54	7.78	10.82
(b) 1979–88	6.07	9.68	16.15	6.30	5.36	10.21	12.16	6.60	10.16
(c) 1988–96	1.94	4.69	9.41	8.58	3.55	2.97	4.61	9.13	11.56

Note: Figures in brackets denote percentage to total.
Source: As in Table 5.3.

American Countries

The United States, which shares more than 77 per cent of the American import of fish in fresh, frozen and chilled form, appears to be shifting her preference for fish product to other product forms during the past 18 years, as is evident from a the gradual fall in her share of the world import. The percentage fall during this period is more than 54 per cent. There is also a substantial fall in the share of this product among all the fish products imported by the United States. The share, which was 45.76 per cent in 1979, has come down to 35.72 per cent in 1996. However, another American country, namely Canada, has expanded her share of the import by 32 per cent during the same period, though Canada's share of world import is not as high as that of the United States. The share of this product among all the fish products imported by Canada has gone up from 24.28 per cent in 1979 to 34.10 per cent in 1996. In the United States the cumulative growth rate during the recent period (1988–96) has fallen, while it has increased in Canada. Canada's general compounded growth in the import of this product group—at around 11 per cent—is also quite healthy.

European Countries

Spain

Among the seven European countries tabled here (which together share more than 84 per cent of European import of this product), all have shown a deceleration in their share of the world import, except Spain whose share of the world market has increased by 83.75 per cent between 1979 and 1996. Spain's cumulative growth rate is highest in all the periods, though there is a fall in growth rate in the last period as compared to the mid-period. There is also an increase, though small, in the share of this fish product among the imports of all fish products in Spain, from 43.07 per cent in 1979 to 44.86 per cent in 1996.

United Kingdom and The Netherlands

The share of the world import market of this product has fallen sharply for the United Kingdom—the fall was about 51 per cent during 1979–96. Its cumulative rate of growth during the recent period has also fallen as compared to the earlier period, which suggests that she is not keeping pace with the growth of world import of this product. The share of this product among imports of all fish products in the United Kingdom has also gone down from 55.44 per cent in 1979 to 38.15 per cent in 1996.

For the Netherlands, there is a marginal fall in the share of the world import of this product. The country has maintained a good growth rate in the import of this product. The share of this product among imports of all fish products in the Netherlands has gone up marginally from 50.36 per cent in 1979 to 52.70 per cent in 1996.

Germany

Germany's share has come down to what Spain has now, though in 1979 its share was more than double that of Spain. During the recent period Germany's rate of growth has increased considerably. Her cumulative rate of growth for the entire period has also been very good. Germany's case is somewhat similar to that of the Netherlands, though the share of import of this product in the total imports of fish products in Germany has gone down marginally from 58.29 per cent in 1979 to 56.71 per cent in 1996.

The remaining countries of Europe have not fared well both in respect of their share of the import market and rate of growth. However, for all these countries the expansion of market in this product has not been very poor, as will be evident from Table 5.7.

TABLE 5.7

Comparative market (import) expansion of fish in fresh, chilled or frozen form in select countries of America and Europe

(per cent)

Country	Expansion of market	
	1979–96	1987–96
United States	98.18	2.03
France	231.62	58.74
Spain	694.83	164.44
Germany	234.80	113.10
United Kingdom	168.32	79.51
Italy	203.25	30.74
Denmark	302.65	52.38
The Netherlands	257.55	121.88
Canada	473.13	129.94
WORLD	333.02	96.97

Asian Countries

The major market for fish in fresh or frozen form lies in Asia and, in a comparatively small way, Africa. Table 5.8 analyses the import status of the major importing countries of this product in Asia.

TABLE 5.8
Import of fish in fresh, chilled or frozen form by major Asian countries

Year	Japan	Korea(R)	Hong Kong	Thailand	Singapore	Malaysia
1979	1083 (19.94)	39 (0.72)	94 (1.73)	2 (3.03)	28 (0.52)	20 (0.37)
1980	853 (15.29)	25 (0.45)	110 (1.97)	11 (0.20)	42 (0.75)	16 (0.29)
1981	1230 (20.29)	50 (0.82)	119 (1.96)	9 (0.15)	61 (1.00)	18 (0.30)
1982	1381 (22.59)	41 (0.67)	130 (2.13)	15 (0.25)	55 (0.90)	27 (0.44)
1983	1391 (24.42)	28 (0.49)	116 (2.03)	29 (0.51)	67 (1.18)	15 (0.26)
1984	1527 (25.38)	48 (0.80)	120 (1.99)	74 (1.23)	77 (1.28)	43 (0.71)
1985	1767 (25.82)	61 (0.89)	117 (1.71)	128 (1.87)	78 (1.14)	47 (0.69)
1986	2539 (27.44)	88 (0.95)	140 (1.51)	263 (2.84)	102 (1.10)	51 (0.55)
1987	3120 (26.13)	174 (1.46)	162 (1.36)	248 (2.08)	117 (0.98)	61 (0.51)
1988	4298 (29.73)	247 (1.71)	205 (1.42)	542 (3.75)	149 (1.03)	63 (0.44)
1989	4103 (28.09)	239 (1.64)	215 (1.48)	679 (4.65)	150 (1.03)	73 (0.50)
1990	4379 (25.85)	246 (1.45)	258 (1.52)	738 (4.36)	150 (0.89)	78 (0.46)
1991	4981 (26.18)	397 (2.09)	307 (1.61)	962 (5.05)	188 (0.99)	89 (0.47)

(Contd.)

Table 5.8 (contd.)

Year	Japan	Korea(R)	Hong Kong	Thailand	Singapore	Malaysia
1992	5656 (29.13)	340 (1.75)	348 (1.79)	817 (4.21)	231 (1.19)	151 (0.78)
1993	6074 (32.90)	360 (1.95)	352 (1.91)	644 (3.48)	234 (1.27)	160 (0.87)
1994	6746 (32.90)	432 (2.11)	415 (2.02)	556 (2.71)	235 (1.15)	180 (0.88)
1995	7142 (31.44)	505 (2.22)	455 (2.00)	500 (2.20)	279 (1.23)	185 (0.85)
1996	7117 (30.26)	615 (2.61)	542 (2.30)	510 (2.17)	245 (1.04)	220 (0.94)

Cumulative growth rate (% p.a.)

(a) 1979–96	11.71	17.61	10.86	38.54	13.61	15.15
(b) 1979–88	16.55	22.76	9.05	86.35	20.41	13.60
(c) 1988–96	6.51	12.08	12.92	(–) 0.76	6.41	16.92

Note: Figures in brackets denote percentage to total.
Source: As in Table 5.3.

The six countries mentioned in Table 5.8 command about 92 per cent of the Asian market for fresh, chilled or frozen form of fish product. Japan is the largest importer among them, commanding about 71 per cent of the Asian market. However, during the last ten years (1988–96) Japan's rate of growth has fallen considerably as compared to the mid period (1979–88). This might indicate a saturation of the market. Part of the Japanese fish market may be switching to other product segments of the market. This is also supported by the fact that though the share of this product in the total import of fish products in Japan has gone up from 27.37 per cent in 1979 to 40 per cent in 1986, it has remained almost stationary till 1996 when the same has gone up marginally to 42.5 per cent. In contrast, the share of prepared and preserved fish products among all import of fish products in Japan has gone up sharply from 6.97 per cent in 1986 to 14.51 per cent in 1996.

The remaining five countries together have presently a share of about 9 per cent in the world imports in fresh, frozen or chilled fish product, which has increased from a negligible 2.65 per cent in 1979. However, except Hong Kong and Malaysia all others have registered a substantial fall in their growth rate during the recent period (1988–96), though, except Thailand, all other countries have increased their share of the world import of this product. A part of the imports by these countries may be re-exported, which shall be examined later. There may also be some shift in preference, as in Japan.

TABLE 5.9

Comparative market (import) expansion of fish in fresh, chilled or frozen form in select countries of Asia

(per cent)

Country	*Expansion of market*	
	1979–96	*1987–96*
Japan	557.16	128.11
Korea(R)	1476.92	253.45
Hong Kong	476.60	234.57
Thailand	25,400.00	105.65
Singapore	775.00	109.40
Malaysia	1000.00	260.65
WORLD	333.02	96.97

Singapore and Malaysia

Singapore and Malaysia deserve special mention. In both countries, the share of import of this product among all the fish product imports has gone up while the share of prepared and preserved fish product has fallen. For Singapore, the share of fresh, frozen or chilled fish product among all the fish product imports has gone up from 29.47 per cent in 1979 to 38.28 per cent in 1996. In the case of Malaysia, the increase is much more, from 25.64 per cent in 1979 to 66.87 per cent in 1996. But in the case of prepared and preserved fish product, the share has gone down from 25.26 per cent in 1979 to 18.60 per cent in 1996 for Singapore, while for Malaysia the fall is from 29.50 per cent in 1979 to a mere 8.80 per cent in 1996. The picture is similar in Korea(R).

The expansion of the market for this product in individual countries of Asia is, however, quite substantial as will be evident from Table 5.9.

IMPORT OF FISH IN PREPARED OR PRESERVED FORM

Import of this product group is the domain of the developed nations of the world, located primarily in America and Europe. Japan is also the major importer of this fish product in the world.

CONTINENT ANALYSIS

Table 5.10 presents import data of this fish product by five continents.

TABLE 5.10
Import of fish in prepared, preserved form by continent

(in million US dollars)

Year	America	Europe	Asia	Africa	Oceania
1979	462 (19.03)	1241 (51.11)	349 (14.37)	245 (10.10)	132 (5.44)
1980	569 (19.62)	1475 (50.86)	429 (14.79)	273 (9.41)	154 (5.31)
1981	618 (20.75)	1389 (46.63)	431 (14.47)	34 9 (11.72)	192 (6.45)
1982	611 (22.75)	1240 (46.17)	471 (17.54)	205 (7.63)	159 (5.92)
1983	657 (24.05)	1340 (49.05)	406 (14.86)	185 (6.77)	144 (5.27)
1984	696 (25.67)	1243 (45.85)	458 (16.89)	145 (5.35)	168 (6.20)
1985	787 (27.06)	1310 (45.05)	510 (17.54)	139 (4.78)	162 (5.57)
1986	898 (36.99)	1898 (50.25)	657 (17.39)	145 (3.84)	179 (4.74)

(Contd.)

Table 5.10 (contd.)

Year	America	Europe	Asia	Africa	Oceania
1987	997 (21.96)	2386 (52.56)	827 (18.22)	137 (3.02)	193 (4.25)
1988	1121 (20.39)	2616 (47.57)	1290 (23.45)	156 (2.84)	230 (4.18)
1989	1121 (20.07)	2772 (49.62)	1221 (21.86)	166 (2.97)	229 (4.10)
1990	1105 (18.28)	3341 (55.28)	1168 (19.32)	167 (2.76)	211 (3.49)
1991	1274 (18.75)	3512 (51.68)	1549 (22.79)	170 (2.50)	229 (3.37)
1992	1241 (17.42)	3729 (52.35)	1722 (24.18)	127 (1.78)	236 (3.31)
1993	1257 (17.83)	3260 (46.24)	2057 (29.18)	138 (1.96)	243 (3.45)
1994	1423 (17.91)	3480 (43.80)	2451 (30.85)	121 (1.52)	251 (3.16)
1995	1491 (16.29)	4015 (43.87)	2893 (31.61)	165 (1.80)	231 (2.52)
1996	1568 (16.28)	4184 (43.45)	3075 (31.93)	134 (1.39)	236 (2.45)

Cumulative growth rate (% p.a.)

(a) 1979–96	7.45	7.41	13.66	(–) 3.49	3.48
(b) 1979–88	10.35	8.64	15.63	(–) 4.89	6.36
(c) 1988–96	4.28	6.05	11.47	(–) 1.88	0.32

Note: Figures in brackets represent the percentage of total world import of this form of fish product.

Source: As in Table 5.3

America and Europe

Despite the fact that the major consumers of fish in prepared and preserved form are from the developed nations of America and Europe, the market shares of these two continents have fallen by about 15 per cent, though there has been a sizeable expansion of the market for this product in America and Europe over the past eighteen years. Although the overall rate of growth in imports of this product has been the same in both the continents, during the recent period (1988–96) the growth rate in America has fallen drastically as compared to that in Europe.

The world market for this product has expanded by 296.58 per cent between 1979 and 1996, while American and European markets have expanded by 239.40 per cent and 237.14 per cent respectively during the same period—their expansion being lower than the world market expansion, which explains their fall in the market share. But the share of this product in total fish products import of America has gone up from 13.71 per cent in 1979 to 17.20 per cent in 1996 which indicates a switch in preference for this product as indicated earlier. However, for Europe the situation is just the opposite. The share of this fish product in total imports has gone down from 24.66 per cent in 1979 to 22.06 per cent in 1996.

Asia

A major market expansion for this product has taken place in Asia, led, once again, by Japan. The share of Asia has more than doubled during 1979–96. The compounded growth rate is also the highest in this region—nearly double that of America and Europe. The size of the Asian market has also grown by 781 per cent as against a world market expansion of 296.58 per cent between 1979 and 1996—more than double of the world market expansion. The share of this product in the total import of fish products in Asia has gone up from 7.46 per cent in 1979 to 13.59 per cent in 1996.

Africa and Oceania

Both the African and Oceanian markets for this product have shrinked considerably since 1979. In the case of Africa, the rate of growth is negative in all the three periods, while for Oceania it has fallen drastically during the recent period (1988–96). The share of this product in the total import of fish products in these two continents has also fallen considerably.

COUNTRY ANALYSIS

Table 5.11 lists some of the major importers of this fish product from the developed world of America and Europe with their quantum. of imports, share in world trade, and growth parameters.

American and European Countries

The United States and Canada together have about 83 per cent share of American market while the six European countries in Table 5.11 together represent 73 per cent of the European market for this product.

It may be seen that except in Italy, there has not been much of an improvement in the share of world import of prepared and preserved fish products in these countries during the last 18 years. In fact, the share has fallen in countries like France, Germany, Canada, and Belgium. The other countries have barely maintained their market share. However, in most of the countries the market size has expanded, though not matching with the world market expansion during 1979–96 and during 1987–96, as will be evident from Table 5.12.

TABLE 5.11

Import of prepared and preserved fish products by major developed countries of America and Europe

(in million US dollars)

Year	United States	United Kingdom	France	Germany	Italy	Denmark	Canada	Belgium
1979	278 (11.44)	238 (9.80)	243 (10.00)	179 (7.37)	73 (3.00)	64 (2.64)	80 (3.29)	124 (5.11)
1980	322 (11.10)	333 (11.48)	280 (9.66)	200 (6.90)	80 (2.76)	80 (2.76)	106 (3.65)	129 (4.45)
1981	379 (12.72)	382 (12.82)	265 (8.90)	164 (5.51)	59 (1.98)	67 (2.25)	94 (3.16)	117 (3.93)
1982	414 (15.41)	283 (10.54)	263 (9.79)	158 (5.88)	58 (2.16)	56 (2.08)	79 (2.94)	108 (4.02)
1983	468 (17.13)	361 (13.12)	270 (9.88)	159 (5.82)	74 (2.71)	60 (2.20)	107 (3.92)	104 (3.81)
1984	493 (18.19)	331 (12.21)	234 (8.63)	152 (5.61)	68 (2.51)	71 (2.62)	120 (4.43)	86 (3.17)
1985	613 (21.08)	334 (11.49)	250 (8.60)	162 (5.57)	94 (3.23)	67 (2.30)	100 (3.44)	87 (2.99)
1986	696 (18.43)	472 (12.50)	349 (9.24)	222 (5.88)	142 (3.76)	123 (3.26)	116 (3.07)	126 (3.34)
1987	738 (16.26)	488 (10.75)	495 (10.90)	267 (5.88)	206 (4.54)	163 (3.59)	153 (3.37)	156 (3.34)
1988	830 (15.09)	586 (10.66)	540 (9.82)	279 (5.07)	221 (4.02)	155 (2.82)	200 (3.64)	161 (2.93)
1989	770 (13.78)	654 (11.71)	535 (9.58)	291 (5.21)	232 (4.15)	176 (3.15)	217 (3.88)	182 (3.26)
1990	810 (13.40)	670 (11.09)	656 (10.86)	416 (6.88)	304 (5.03)	205 (3.39)	174 (2.88)	223 (3.69)

(Contd.)

Table 5.11 (contd.)

Year	United States	United Kingdom	France	Germany	Italy	Denmark	Canada	Belgium
1991	972 (14.30)	720 (10.59)	625 (9.20)	491 (7.22)	370 (5.44)	165 (2.43)	188 (2.77)	216 (3.18)
1992	898 (12.60)	765 (10.74)	625 (8.77)	500 (7.02)	427 (5.99)	205 (2.88)	188 (2.64)	222 (3.12)
1993	851 (12.07)	715 (10.14)	547 (7.76)	405 (5.74)	348 (4.94)	192 (2.72)	229 (3.25)	192 (2.72)
1994	952 (11.98)	630 (7.93)	604 (7.60)	511 (6.43)	360 (4.53)	229 (2.88)	238 (2.99)	207 (2.61)
1995	970 (10.60)	794 (8.68)	780 (8.52)	546 (5.96)	397 (4.34)	278 (3.04)	255 (2.79)	241 (2.63)
1996	1076 (11.17)	851 (8.84)	774 (8.04)	575 (5.97)	430 (4.47)	278 (2.89)	233 (2.42)	228 (2.37)

Cumulative growth rate (% p.a.)

	United States	United Kingdom	France	Germany	Italy	Denmark	Canada	Belgium
(a) 1979–96	8.29	7.78	7.05	7.11	10.99	9.02	6.49	3.65
(b) 1979–88	12.92	10.53	9.28	5.06	13.10	10.33	10.72	2.94
(c) 1988–96	3.30	4.77	4.60	9.46	8.68	7.58	1.93	4.45

Note: Figures in brackets represent percentage share world imports in this form of fish product.
Source: As in Table 5.3.

United States

The United States' share of world import of this fish product went up steeply during the mid-1980s. But from the late 1980s, the share has declined continuously to finally touch the share level of 1979. However, the share of this product in the total import of fish products in the United States has increased from 10.04 per cent in 1979 to 15.31 per cent in 1996. These findings indicate that although there has been a distinct shift in preference for this form of fish product in the United States, not much additional demand is created in US market. United States does not hold any high prospect for this form of fish product in future. Canada's case is similar to that of the United States, except that the fall in her market share is comparatively larger than that of the United States.

TABLE 5.12

Comparative market (import) expansion of prepared and preserved fish products in select countries of America and Europe

(per cent)

Country	Expansion of market	
	1979–96	*1987–96*
United States	287.05	45.80
United Kingdom	257.56	74.38
France	218.52	56.36
Germany	221.23	115.36
Italy	489.04	108.74
Denmark	334.38	70.55
Canada	191.25	52.29
Belgium	83.87	46.15
WORLD	296.58	112.09

United Kingdom

The import characteristics of this product for the United Kingdom are closely similar to that of the United States and Canada. Her share of world import also increased during the mid-1980s (though, not as high as that of the United States), but then began declining. In 1996, it could barely recover her lost position. However, the share of this product in the total import of fish products in the United Kingdom has gone up from 34.54 per cent in 1979 to 40.18 per cent in 1996. Here also, after the initial flush of enthusiasm the charm is rather dying down. The demand for this product in United Kingdom is stagnating.

Denmark

Denmark's share of world import in this product has remained almost stationary during the past eighteen years, though the rate of growth has fallen during the recent period. Her total market expansion during 1979–96 is larger than the world market expansion which has, however, gone down during the last ten years. The share of this product in the total import of fish products in Denmark has decreased marginally from 24.24 per cent in 1979 to 22.36 per cent in 1996.

Germany

Germany's share of the world import of this fish product has fallen marginally during the last 18 years, as has the share of this product in the total import of fish products in the country (from 25.57 per cent in 1979 to 24.29 per cent in 1996). However, unlike many of the European countries there is a spurt in the rate of growth of import during the recent period (1988–96). This is also reflected in the market expansion rate during the past ten years.

France

There has been a decline in France's share of world import of this product by 2 percentage points. There is also a decline in her growth rate during the recent period. But the share of this product in the total imports of fish product in France has remained at around 24 per cent during the entire period.

Italy

The import performance of this country in this fish product is the best among the European countries considered here, by almost all the parameters. There is an increase in the share of the world import. The market expansion rate is close to that of the world expansion rate in this product. The share of this product in total imports of fish products has also gone up from 10.85 per cent in 1979 to 17.09 per cent in 1996. It appears that Italy provides a growing market in prepared and preserved fish products.

Country-analyses of America and Europe made above suggest that except Italy, not much additional demand for this fish product is being created in these markets, and that the initial enthusiasm may be tapering off.

Asian Countries

In contrast to the developed countries of Europe and America, Japan has increased her share of the import of prepared and preserved fish products substantially. This is also the case with some other advanced developing nations of Asia, as will be evident from Table 5.13.

TABLE 5.13

Import of prepared and preserved fish products by major countries of Asia

(in million US dollars)

Year	Japan	Hong Kong	Singapore	Malaysia
1979	176 (7.25)	22 (0.91)	24 (0.99)	23 (0.95)
1980	183 (6.31)	30 (1.03)	34 (1.17)	31 (1.07)
1981	184 (6.18)	34 (1.14)	32 (1.07)	30 (1.01)
1982	214 (7.97)	32 (1.19)	33 (1.23)	34 (1.27)
1983	216 (7.91)	36 (1.32)	29 (1.06)	30 (1.10)
1984	258 (9.52)	31 (1.14)	33 (1.22)	35 (1.29)
1985	308 (10.59)	38 (1.31)	25 (0.86)	35 (1.20)
1986	448 (11.86)	51 (1.35)	32 (0.85)	37 (0.98)
1987	600 (13.22)	66 (1.45)	36 (0.79)	38 (0.84)
1988	962 (17.49)	72 (1.31)	43 (0.78)	34 (0.62)
1989	951 (17.02)	80 (1.43)	40 (0.72)	40 (0.72)
1990	916 (15.16)	84 (1.39)	39 (0.65)	22 (0.36)
1991	1226 (18.04)	102 (1.50)	49 (0.72)	28 (0.41)
1992	1370 (19.23)	127 (1.78)	59 (0.83)	27 (0.38)
1993	1680 (23.83)	131 (1.86)	69 (0.98)	31 (0.44)
1994	1960 (24.67)	161 (2.03)	87 (1.09)	34 (0.43)
1995	2336 (25.52)	141 (1.54)	102 (1.11)	33 (0.36)
1996	2419 (25.12)	191 (1.98)	119 (1.24)	29 (0.30)

Cumulative growth rate (% p.a.)

(a) 1979–96	16.67	13.56	9.88	1.37
(b) 1979–88	20.77	14.08	6.69	4.44
(c) 1988–96	12.22	12.97	13.57	(–) 1.97

Note: Figures in brackets represent percentage share of total world import in this fish product.
Source: As in Table 5.3.

The four Asian countries mentioned in Table 5.13 together represent nearly 90 per cent of the import of this product in Asia.

Japan

Japan's share of world import of this fish product has increased by more than three times between 1979 and 1996. It is significant that presently Japan alone has a share of more than 25 per cent of the world import of prepared and preserved fish products, which is more than double that of the United States and more than half that of the entire Europe. While the world market of this product grew by about 297 per cent between 1979 and 1996, Japan's market expanded by about 1247 per cent during the same period, though during the recent period there is a fall in the growth rate of import of this product in the country.

Japan is the leading importer of both fresh, chilled or frozen form of fish products and prepared and preserved form of fish products in the world. The expansion of the latter market is partly due to a shift from the former market. The share of this product in Japan's total import of fish products has gone up from 4.45 per cent in 1979 to 14.50 per cent in 1996. Japan also exports a small quantity of this product.

Hong Kong, Singapore and Malaysia

Though both Hong Kong and Singapore have a small share in the world import of this product, their rates of growth are quite significant. The share of Malaysia in the world import of this product has fallen, as also her rate of growth.

TABLE 5.14

Comparative market (import) expansion of preserved and prepared fish products in select Asian countries between 1979 and 1996

(per cent)

Country	Expansion of market	
	1979–96	*1987–96*
Japan	1274.43	303.17
Hong Kong	768.18	189.39
Singapore	395.83	230.56
Malaysia	26.09	(–) 23.68
WORLD	296.58	112.09

Some of these countries are re-exporters of this product, as we shall see later. The comparative rates of expansion of the market of this product in Asian countries are calculated in Table 5.14.

IMPORT OF CRUSTACEANS, MOLLUSCS ETC. IN FRESH, FROZEN OR CHILLED FORM

Asia accounts for nearly 50 per cent of the world import of shellfish (crustaceans, molluscs etc.) in fresh, frozen or chilled form, as will be evident from Table 5.15.

TABLE 5.15
Import of shellfish (crustaceans, molluscs, etc.) in fresh, frozen or chilled form by continent

(in million US dollars)

Year	America	Europe	Asia	Africa	Oceania
1979	1314 (27.70)	897 (18.91)	2490 (52.50)	16 (0.34)	25 (0.51)
1980	1310 (29.13)	1108 (24.64)	2025 (45.03)	19 (0.42)	34 (0.76)
1981	1448 (30.87)	976 (20.81)	2196 (46.82)	30 (0.64)	40 (0.86)
1982	1683 (32.64)	1022 (19.82)	2385 (46.26)	22 (0.43)	45 (0.87)
1983	2120 (37.52)	1095 (19.38)	2358 (41.73)	23 (0.41)	54 (0.96)
1984	2124 (37.24)	1064 (18.65)	2441 (42.79)	23 (0.40)	53 (0.93)
1985	2119 (35.26)	1159 (19.28)	2658 (44.22)	21 (0.35)	55 (0.92)
1986	2476 (30.55)	1851 (22.83)	3700 (45.65)	15 (0.19)	64 (0.79)
1987	2827 (27.96)	2589 (25.60)	4585 (45.34)	17 (0.17)	95 (0.94)
1988	2781 (24.30)	2814 (24.59)	5734 (50.10)	26 (0.23)	86 (0.75)
1989	2772 (24.27)	3057 (26.76)	5456 (47.76)	24 (0.21)	97 (0.85)
1990	2671 (22.08)	3492 (28.87)	5758 (47.61)	27 (0.22)	114 (0.94)
1991	2865 (21.66)	3892 (29.43)	6328 (47.85)	22 (0.17)	110 (0.83)
1992	3176 (23.42)	3869 (28.53)	6378 (47.04)	34 (0.25)	94 (0.69)
1993	3374 (23.89)	3558 (25.19)	7026 (49.75)	29 (0.21)	102 (0.72)
1994	4011 (23.82)	4118 (24.46)	8455 (50.21)	33 (0.20)	150 (0.89)
1995	3948 (21.63)	4817 (26.39)	9196 (50.38)	43 (0.24)	164 (0.90)
1996	3833 (22.03)	4686 (26.93)	8588 (49.36)	34 (0.20)	164 (0.94)

(Contd.)

Table 5.15 (contd.)

Cumulative growth rate (% p.a.)

(a) 1979–96	6.50	10.21	7.55	4.53	11.70
(b) 1979–88	8.69	13.55	9.71	5.54	14.71
c) 1988–96	4.09	6.58	5.18	3.41	8.40

Note: Figures in brackets represent percentage share of total world import of this fish product.

Source: As in Table 5.3.

CONTINENT ANALYSIS

America

The rate of increase in the import of crustaceans in fresh, frozen or chilled form is on the decline in America, as is evident not only from a fall in her share of the world import but also from the declining trend of compounded growth rate during the recent period (1988–96), which has fallen by more than half of the rate of growth in the earlier period. While the world market for this product has expanded by about 267 per cent, in America the expansion is only about 192 per cent, resulting in a fall in her share of the market. Although there is a moderate increase in the share of this product in the total import of fish products in America (from 40 per cent in 1979 to 42 per cent in 1996) and although this product has also displaced the number one position of fresh, frozen or chilled fish product, the preference towards prepared and preserved form of fish is greater.

Europe

As compared to America, Europe has fared better both in terms of share and growth in imports of this fish product, though during the recent period (1988–96) the rate of growth has fallen (similar to that in America and all the other continents). The market expansion of this product in Europe is nearly 422.5 per cent between 1979 and 1996, which is much higher than the expansion of world trade in this product. It appears that though the market share of Europe in fresh, frozen or chilled form of fish has declined, it has considerably increased its share in crustaceans, molluscs etc. in the same form. The American market, however, behaved uniformly in both cases, though the fall in the market share of fish in fresh, frozen or chilled form is much higher than that for crustaceans etc. in the same form.

It has been seen earlier that the shares of both fresh, frozen or chilled fish and prepared or preserved form of fish in the total import of fish products in Europe have decreased during the past 18 years. This fall appears to be compensated by a rise in the share of shellfish in fresh, frozen or chilled form in the import basket of Europe. The rise in share is from 17.82 per cent in 1979 to 24.71 per cent in 1996. By all indications Europe provides a growing market for shellfish.

Asia

In absolute quantitative terms, Asia's import of crustaceans etc. is the highest in the world, as also its market share—more than that of America and Europe together. Asia's market share has, however, not increased since 1979—rather, there has been a fall. The compounded growth rate of import of this product has also fallen during the recent period, which, however, is common for all the continents. The expansion of the Asian market in crustaceans etc. in fresh, frozen or chilled form is also not matching with the world expansion. The size of the Asian market has expanded by about 245 per cent between 1979 and 1996 as compared to a world market expansion of 267 per cent. The rate of expansion is much below that of Europe.

Although Asia still holds the largest share of world import of this product, the import performance of the continent is not encouraging. The reason behind this is that the share of shellfish in fresh, frozen or chilled form in the total fish import basket of Asia is on a sharp decline—from 53.27 per cent in 1979 to 37.95 per cent in 1996. This product, which held first position in the share of import basket of fish product in 1979, is now relegated to third position by fresh, frozen or chilled form and by prepared and preserved form of fish products.

Oceania and Africa

In Oceania, the market of this product is still small in size but the rate of expansion is very high, about 556 per cent within a span of 18 years. The share of this product in the total import basket of fish products has also registered a phenomenal increase during the period, from 9.96 per cent in 1979 to 27.29 per cent in 1996. Oceania thus provides a growing market for this product.

The African market is small in size, and its share has also fallen consistently over past 18 years. The share of this product in the fish import basket of Africa has increased marginally from 3.10 per cent in 1979 to 4.20 per cent in 1996. The import market in Africa has shifted heavily towards fresh, frozen or chilled fish products.

Crustaceans etc. are among high value aqua products aimed at the upper end of the market. Countries with high gross national products or per capita incomes are obviously major consumers of this product. Due to low incomes, many of the developing countries which are major producers of these products cannot also afford to consume them. These products are produced there principally for the export market.

COUNTRY ANALYSIS

Table 5.16 presents the import data of this product by major importers of America and Europe.

TABLE 5.16
Import of crustaceans, molluscs etc. by major importers of America and Europe

(in million US dollars)

Year	United States	Spain	France	Italy	Canada
1979	1157 (24.39)	163 (3.44)	274 (5.78)	122 (2.57)	124 (2.61)
1980	1135 (25.24)	225 (5.00)	314 (6.98)	162 (3.60)	112 (2.49)
1981	1243 (26.50)	169 (3.60)	293 (6.24)	135 (2.88)	120 (2.56)
1982	1479 (28.69)	189 (3.67)	292 (5.66)	159 (3.08)	131 (2.54)
1983	1883 (33.33)	177 (3.13)	302 (5.35)	167 (2.96)	155 (2.74)
1984	1910 (33.49)	189 (3.31)	269 (4.72)	172 (3.02)	167 (2.93)
1985	1917 (31.90)	164 (2.73)	286 (4.76)	248 (4.13)	160 (2.66)
1986	2249 (27.75)	314 (3.87)	454 (5.60)	358 (4.42)	194 (2.39)
1987	2577 (25.48)	593 (5.86)	569 (5.63)	463 (4.58)	214 (2.12)
1988	2528 (22.09)	627 (5.48)	599 (5.23)	545 (4.76)	220 (1.92)
1989	2511 (21.98)	799 (6.99)	612 (5.36)	582 (5.09)	231 (2.02)
1990	2395 (19.80)	935 (7.73)	712 (5.89)	639 (5.28)	238 (1.97)
1991	2562 (19.37)	1000 (7.56)	781 (5.90)	723 (5.47)	258 (1.95)
1992	2831 (20.88)	1135 (8.37)	749 (5.52)	702 (5.18)	253 (1.87)
1993	3006 (21.29)	970 (6.87)	694 (4.91)	592 (4.19)	278 (1.97)
1994	3613 (21.46)	1090 (6.47)	791 (4.70)	672 (3.99)	308 (1.83)
1995	3535 (19.37)	1354 (7.42)	918 (5.03)	772 (4.23)	322 (1.76)
1996	3309 (19.02)	1287 (7.40)	891 (5.12)	802 (4.61)	445 (2.56)

Cumulative growth rate (% p.a.)

(a) 1979–96	6.38	12.92	7.18	11.71	7.81
(b) 1979–88	9.97	16.15	9.08	18.09	6.58
(c) 1988–96	3.42	9.41	5.09	4.95	9.20

(Contd.)

Table 5.16 (contd.)

Year	Denmark	Belgium	United Kingdom	The Netherlands	Germany
1979	30 (0.63)	73 (1.54)	62 (1.31)	40 (0.84)	46 (0.97)
1980	53 (1.79)	83 (1.85)	67 (1.49)	54 (1.20)	54 (1.20)
1981	56 (1.19)	64 (1.36)	70 (1.49)	46 (0.98)	46 (0.98)
1982	53 (1.03)	58 (1.12)	79 (1.53)	37 (0.72)	49 (0.95)
1983	72 (1.27)	67 (1.19)	89 (1.58)	40 (0.71)	54 (0.96)
1984	78 (1.37)	61 (1.07)	92 (1.61	34 (0.60)	48 (0.84)
1985	97 (1.61)	61 (1.02)	89 (1.48)	35 (0.58)	50 (0.83)
1986	148 (1.83)	103 (1.27)	144 (1.78)	48 (0.59)	82 (1.01)
1987	222 (2.20)	129 (1.28)	179 (1.77)	72 (0.71)	99 (0.98)
1988	225 (1.97)	154 (1.35)	192 (1.68)	81 (0.71)	123 (1.07)
1989	235 (2.06)	158 (1.38)	183 (1.60)	77 (0.67)	126 (1.10)
1990	242 (1.83)	191 (1.44)	192 (1.45)	86 (0.65)	152 (1.15)
1991	254 (1.87)	206 (1.52)	181 (1.34)	129 (0.95)	169 (1.25)
1992	201 (1.42)	219 (1.55)	154 (1.09)	135 (0.96)	179 (1.27)
1993	220 (1.31)	220 (1.31)	165 (0.98)	159 (0.94)	173 (1.03)
1994	236 (1.29)	287 (1.57)	239 (1.31)	178 (0.98)	222 (1.22)
1995	262 (1.44)	325 (1.78)	253 (1.39)	231 (1.27)	227 (1.24)
1996	291 (1.67)	276 (1.59)	227 (1.30)	215 (1.24)	201 (1.16)

Cumulative growth rate (% p.a.)

(a) 1979–96	14.30	8.14	7.93	10.40	9.06
(b) 1979–88	25.09	8.65	13.98	8.16	11.55
(c) 1988–96	3.27	7.57	2.12	12.98	6.33

Note: Figures in brackets represent percentage share of total world import of this fish product.

Source: As in Table 5.3.

American Countries

United States and Canada

United States, being the dominant importer of crustaceans, molluscs, etc. in fresh, frozen or chilled form, also exhibits the same characteristics of falling market share and considerable deceleration in the growth of imports during the recent period (1988–96). However, this product holds the major share in the import of all fish products in the United States. The share has also gone up from 41.78 per cent in 1979 to 47.08 per cent in 1996. This has eaten into the share of fresh, frozen or chilled fish product in the import basket of fish in the United States. But considering all the findings together, it can be said that the United States no longer shows any signs of growing characteristics of the market.

Canada's market share also exhibits a downward trend. But for the year 1996, the share of Canada can be said to have fallen between 1979 and 1996. However, in percentage terms, Canada's market for this product has expanded by 285 per cent (if the year 1996 is included), as compared to only 186 per cent in the United States. The share of this product in Canada's import of all fish products has gone down from 44.93 per cent in 1979 to 39.52 per cent in 1996. Like the United States, Canada has also ceased to exhibit growing characteristics of the market. Canada and United States together account for about 98 per cent of the American market of this product.

European Countries

All the eight countries of Europe mentioned in Table 5.16 have increased their share of import of crustaceans in fresh, frozen or chilled form, except France, Belgium, and the United Kingdom who have just been able to maintain their market share.

Spain

Spain has the highest share of world import of this fish product in the whole of Europe. The share has also shown an increasing trend during the past 18 years. The rate of growth in imports is quite high. The share of this product is the highest among the total import of fish products in Spain. There is also a moderate increase in this share during the past 18 years, from 40.35 per cent in 1979 to 41.75 per cent in 1996. We can say, therefore, that Spain will continue to provide a stable market for this product.

Denmark

This country had a small share (0.63 per cent) of the world import of this product in 1979, which has increased to 1.67 per cent in 1996. Percentage-wise, the increase is about 165 per cent, which is, however, below the world expansion rate of about 267 per cent during the period under study. This is also evident in the decline in the growth rate of import during the recent period (1988–96).

The largest share of Denmark's import basket of fish is held by fish in fresh, frozen or chilled form, though the share has fallen by about 8 percentage points between 1979 and 1996. Percentage-wise, this fall in share is more than offset by an increase in the share of shellfish in fresh, frozen or chilled form from 11.36 per cent in 1979 to 23.41 per cent in 1996. This suggests that there is a steady rise in the demand for this product in Denmark and, hence, the country provides a good market for the future.

The Netherlands

This country's share of world import of shellfish in fresh, frozen or chilled form has also increased, though moderately, during the past 18 years. Her rate of growth in import during the recent period is also quite impressive. The share of this product in the import basket of fish products has also increased from 14.50 per cent in 1979 to 22.80 per cent in 1996. Netherlands thus provides a growing market for this fish product.

Italy

This predominantly fish eating country's share of world import of this product has gone up by about 80 per cent between 1979 and 1996. The rate of growth in import was very high during 1979 and 1988, but during the later period 1988–96 it has declined substantially. Like the United Kingdom and Germany, there is a ten percentage points loss of preference each for fish in fresh, frozen or chilled form and salted, smoked or dried form in the total import of fish products in Italy. Their losses are the gains of the remaining two fish products. The share of crustaceans, molluscs etc. in fresh, frozen or chilled form has increased from 18.13 per cent in 1979 to 31.88 per cent in 1996 in the total import basket of fish products in Italy. Considering all these findings it can be said that Italy provides a growing market for this product.

Belgium, France and United Kingdom

These three countries have maintained their share of the world import of this product consistently throughout the period under study. There is not much of an increase or decrease in the share which indicates that they have moved along the growth of world expansion in the import of this product. The nature of the growth pattern of import is also similar in these three countries. The share of crustaceans, molluscs etc. in fresh, frozen or chilled form in the total import of fish products has also remained constant at around 10 per cent for the United Kingdom and 28 per cent for France, while in case of Belgium it has gone up from 21 per cent in 1979 to 33 per cent in 1996. These three countries together provide a stable market for this fish product.

Germany

Germany's share of the world import of this product has increased moderately. The share of this product in the country's import platter has increased from 6.57 per cent in 1979 to 8.50 per cent in 1996. Germany provides a growing market for this product.

The eight countries of Europe have nearly 90 per cent share of the European market in this fish product. The comparative market expansion rates of this product in developed countries of America and Europe are summarized in Table 5.17.

TABLE 5.17

Comparative market (import) expansion of crustaceans,
molluscs etc. in fresh, frozen or chilled form in select
developed countries of America and Europe

(per cent)

Country	Expansion of market	
	1979–96	*1987–96*
United States	186.00	28.40
Spain	689.57	117.03
France	225.18	56.59
Italy	557.38	73.22
Canada	258.87	107.94
Denmark	870.00	31.08
Belgium	278.08	113.95
United Kingdom	266.13	26.82
Netherlands	437.50	198.61
Germany	336.96	103.03
WORLD	266.84	72.06

Asian Countries

Table 5.18 presents import data of crustaceans, molluscs etc. by major importing countries of Asia.

TABLE 5.18
*Import of crustaceans, molluscs etc. in fresh, chilled or
frozen form by major importers of Asia*

(in million US dollars)

Year	Japan	Hong Kong	Korea (R)	Singapore	Thailand
1979	2265 (47.75)	147 (3.10)	15 (0.32)	26 (0.55)	15 (0.32)
1980	1785 (39.69)	167 (3.71)	7 (0.16)	26 (0.58)	8 (0.18)
1981	1964 (41.88)	149 (3.18)	3 (0.06)	32 (0.68)	9 (0.19)
1982	2053 (39.82)	229 (4.44)	11 (0.21)	47 (0.91)	8 (0.16)
1983	2028 (35.89)	217 (3.84)	8 (0.14)	65 (1.15)	10 (0.18)
1984	2099 (36.80)	217 (3.80)	12 (0.21)	71 (1.24)	9 (0.16)
1985	2285 (38.02)	242 (4.03)	18 (0.30)	68 (1.13)	6 (0.01)
1986	3197 (39.44)	340 (4.19)	21 (0.26)	88 (1.09)	15 (0.19)
1987	3947 (39.03)	433 (4.28)	24 (0.24)	119 (1.18)	13 (0.13)
1988	4841 (42.30)	580 (5.07)	45 (0.39)	129 (1.13)	12 (0.10)
1989	4651 (40.72)	529 (4.63)	49 (0.43)	133 (1.16)	19 (0.17)
1990	4849 (40.09)	585 (4.84)	79 (0.65)	136 (1.12)	22 (0.18)
1991	5283 (39.94)	605 (4.57)	124 (0.94)	177 (1.34)	43 (0.33)
1992	5190 (38.28)	641 (4.73)	106 (0.78)	197 (1.45)	64 (0.47)
1993	5783 (40.95)	641 (4.54)	105 (0.74)	208 (1.47)	80 (0.57)
1994	6773 (40.22)	799 (4.74)	190 (1.13)	244 (1.45)	117 (0.69)
1995	7363 (40.34)	885 (4.85)	184 (1.01)	226 (1.24)	182 (1.00)
1996	6686 (38.43)	869 (4.99)	254 (1.46)	225 (1.29)	185 (1.06)

Cumulative growth rate (% p.a.)

(a) 1979–96	6.57	11.02	18.11	13.54	15.93
(b) 1979–88	8.81	16.48	12.98	19.48	(–) 2.45
(c) 1988–96	4.12	5.18	24.15	7.20	40.77

Note: Figures in brackets represent percentage share of total world import of this fish product
Source: As in Table 5.3.

Japan

Among the Asian countries, Japan is the largest importer of crustaceans, molluscs etc. in fresh, chilled or frozen form. Her import is double that of the United States and is substantially higher than even the American or European continents. In Asia, Japan has about 78 per cent of the market of this product. However, Japan's share of the market has fallen by about 20 per cent between 1979 and 1996. Her rate of growth has also fallen during the recent period. The expansion of market during this period has been about 195 per cent, as compared to a world market expansion of 267 per cent in this product.

The share of this fish product in Japan's total import of fish products has gone down from 57.24 per cent in 1979 to 40.12 per cent in 1996. This fall is in favour of a rise in the share of fish in fresh, frozen or chilled form and prepared and preserved form. Japan is still a huge market for shellfish in fresh, frozen and chilled form, but it is not a growing market.

Other Countries

The remaining four countries of Asia mentioned in Table 5.18 have, however, increased their share of import of this fish product. Of them, the rate of growth has increased considerably for Thailand and Korea(R) during the recent period, while for others, the rate of growth has fallen. The decline in Japan's market share is a gain for these countries.

TABLE 5.19

Comparative market (import) expansion of crustaceans, molluscs etc.in fresh, frozen or chilled form in select Asian countries

(per cent)

Country	Expansion of market	
	1979–96	1987–96
Japan	195.19	69.39
Hong Kong	491.16	100.69
Korea(R)	1593.33	958.33
Singapore	765.38	89.08
Thailand	1133.33	1323.08
WORLD	266.84	72.06

Among these four countries of Asia, Korea(R) and Singapore are re-exporters of this fish product and Thailand is a major exporter. The share of this fish product in Korea(R)'s total import of fish product remained constant at around 28 per cent during the past 18 years while in Singapore the share has increased from 27.37 per cent in 1979 to 35.16 per cent in 1996.

Hong Kong's share of the world import of this fish product has increased by about 61 per cent between 1979 and 1996. Her rate of growth in import of this product was impressive during 1979–88, but in the latter period 1988–96 the rate has declined. The share of this product in the total import of fish products in Hong Kong is the highest, and has remained more or less constant at around 47 per cent during the entire period. Hong Kong shows the characteristics of a steady market for this fish product.

The comparative market expansion rate of the Asian countries during 1979–96 and 1987–96 are calculated in Table 5.19.

IMPORT OF FISH IN DRIED, SALTED OR SMOKED FORM

The oldest methods of fish preservations are in dried, salted or smoked form. This method still continues, and though the share of this group in the total disposition of fish catch has fallen, as we have seen before, due to the advancement in freezing technology, it still commands about 5.5 per cent of the world fish imports.

CONTINENT ANALYSIS

Table 5.20 shows the continent-wise imports of this product group.

TABLE 5.20
Import of fish in salted, dried and smoked form by continent

(in million US dollars)

Year	America	Europe	Asia	Africa	Oceania
1979	177 (13.88)	437 (34.27)	512 (40.16)	111 (8.71)	11 (0.86)
1980	179 (14.44)	554 (44.68)	301 (24.27)	194 (15.65)	11 (0.89)
1981	193 (14.40)	472 (35.22)	375 (27.29)	287 (21.42)	14 (1.05)
1982	192 (16.67)	443 (38.45)	386 (33.51)	117 (10.16)	15 (1.30)
1983	155 (13.95)	466 (41.94)	362 (32.58)	118 (10.62)	11 (0.99)
1984	158 (15.70)	448 (44.53)	330 (32.80)	57 (5.67)	12 (1.19)
1985	164 (14.32)	549 (47.95)	371 (32.40)	48 (4.19)	13 (1.14)
1986	220 (15.66)	727 (51.74)	382 (27.19)	64 (4.56)	13 (0.93)

(Contd.)

Table 5.20 (contd.)

Year	America	Europe	Asia	Africa	Oceania
1987	213 (11.23)	1079 (56.91)	518 (27.32)	70 (3.69)	16 (0.84)
1988	209 (10.05)	1108 (53.27)	629 (30.24)	54 (2.60)	17 (0.82)
1989	274 (13.75)	1027 (51.53)	567 (28.45)	52 (2.61)	15 (0.75)
1990	278 (11.65)	1362 (57.08)	614 (25.73)	50 (2.10)	15 (0.63)
1991	299 (11.89)	1478 (58.77)	654 (26.00)	53 (2.11)	14 (0.56)
1992	293 (11.50)	1408 (55.26)	773 (30.34)	52 (2.05)	14 (0.55)
1993	313 (13.90)	1097 (48.73)	777 (34.52)	38 (1.69)	15 (0.67)
1994	353 (14.07)	1227 (48.92)	807 (32.18)	51 (2.03)	16 (0.64)
1995	463 (15.90)	1402 (48.15)	918 (31.52)	46 (1.58)	17 (0.58)
1996	457 (15.69)	1425 (48.92)	921 (31.62)	41 (1.41)	18 (0.62)

Cumulative growth rate (% p.a.)

(a) 1979–96	5.74	7.20	3.51	(–) 5.69	2.94
(b) 1979–88	1.86	10.89	2.31	(–) 7.69	4.96
(c) 1988–96	10.27	3.20	4.88	(–) 3.38	0.72

Note: Figures in brackets represent percentage share of total world import of this product.
Source: As in Table 5.3.

America and Europe

It is interesting to note that both America and Europe, the home of the developed nations, have increased their share of import of fish in dried, salted and smoked form while Asia, Africa, and also Oceania, which together had the largest share of import in 1979, had lost their share considerably—more than 32 per cent by 1996.

As seen earlier, both America and Europe had reduced their share of world fish import in both the two forms, namely fresh, frozen or chilled form and prepared or preserved form. It is likely that this deceleration in the share of these two fish markets has not affected the preference of the fish eating people of these two continents towards the traditional dried, salted or smoked form of fish. The reason behind the import market steadiness of this fish product in America and Europe is that the share of this product in the total import of fish products in America and Europe has remained constant at around 5 per cent and 8 per cent, respectively, during the last 18 years. Both these two continents provide a steady market for this fish product, though in some countries, particularly America, the import markets are stagnating.

Although volume-wise the import of this fish form is smallest among the four groups, the above finding needs further investigation which might give new strategic direction in the global marketing of fish products. The monetary size of the market of this product, about US$ 3000 million, cannot also be ignored.

The compounded growth rate in import of this product has fallen in Europe during the recent period (1988–96), but it has increased considerably in America during the same period.

Asia, Africa, and Oceania

Asia's share of the world import of this product has fallen from about 40 per cent in 1979 to about 32 per cent, though the rate of growth has increased during the recent period as compared to the earlier period. Africa has not only lost its share of import, its growth rate has also been consistently negative. Oceania's import is small; its share has also fallen as well as the growth rate during the recent period.

The share of this product in the total import of fish products in these three continents has also gone down substantially. For Asia the fall is from 10.94 per cent in 1979 to 4.07 per cent in 1996 while for Africa the fall is from 21.40 per cent to 5.10 per cent. In Oceania the fall is rather marginal—from 4.40 per cent in 1979 to 3 per cent in 1996. These three continents are showing signs of a declining market.

Country Analysis

Table 5.21 analyses the import data of major importers of dried, salted or smoked fish product from America and Europe.

American and European Countries

Portugal

Of the seven major importing countries listed in Table 5.21, Portugal tops the list, followed by Italy, in respect of the share of import of fish in dried, salted or smoked form, but in both the two countries cumulative growth rate has become negative during the recent period (1988–96).

It is interesting to note here that the share of this product in the total import of fish products in Portugal has gone up substantially from 8.48 per cent in 1979 to 40 per cent in 1996 while the share of fish in fresh, frozen or chilled form has gone down from 85.50 per cent in 1979 to 36 per cent

Table 5.21
Import of dried, salted or smoked fish by major importers of America and Europe.

(in million US dollars)

Year	Portugal	Italy	Germany	Brazil	Spain	United States	France
1979	14 (1.10)	140 (10.98)	67 (5.25)	50 (3.92)	58 (4.55)	67 (5.25)	44 (3.45)
1980	15 (1.21)	168 (13.55)	72 (5.81)	53 (4.27)	93 (7.50)	61 (4.92)	55 (4.44)
1981	29 (2.16)	130 (9.70)	58 (4.33)	41 (3.06)	74 (5.52)	83 (6.19)	44 (3.28)
1982	34 (2.95)	136 (11.80)	55 (4.77)	48 (4.17)	62 (5.38)	77 (6.68)	39 (3.38)
1983	113 (10.17)	121 (10.89)	55 (4.95)	26 (2.34)	39 (3.51)	71 (6.39)	36 (3.24)
1984	106 (10.54)	127 (12.62)	49 (4.87)	28 (2.78)	37 (3.68)	74 (7.36)	35 (3.48)
1985	160 (13.97)	152 (13.28)	51 (4.45)	36 (3.14)	41 (3.58)	74 (6.46)	42 (3.67)
1986	186 (13.24)	178 (12.67)	72 (5.12)	81 (5.77)	77 (5.48)	83 (5.91)	60 (4.27)
1987	311 (16.40)	281 (14.82)	94 (4.96)	56 (2.95)	108 (5.70)	96 (5.06)	83 (4.38)
1988	334 (16.06)	280 (13.46)	102 (4.90)	48 (2.31)	130 (6.25)	92 (4.42)	85 (4.09)
1989	255 (12.79)	275 (13.80)	97 (4.87)	85 (4.26)	125 (6.27)	108 (5.42)	84 (4.21)
1990	379 (15.88)	314 (13.16)	144 (6.04)	80 (3.35)	180 (7.54)	119 (4.99)	94 (3.94)

(Contd.)

Table 5.21 (contd.)

Year	Portugal	Italy	Germany	Brazil	Spain	United States	France
1991	417 (16.58)	358 (14.23)	163 (6.48)	94 (3.74)	186 (7.40)	126 (5.01)	110 (4.37)
1992	381 (14.95)	307 (12.05)	182 (7.14)	69 (2.71)	196 (7.69)	135 (5.30)	109 (4.28)
1993	278 (12.35)	240 (10.66)	132 (5.86)	94 (4.18)	150 (6.67)	123 (5.46)	84 (3.73)
1994	285 (11.36)	266 (10.61)	176 (7.02)	124 (4.94)	158 (6.30)	124 (4.94)	103 (4.11)
1995	331 (11.37)	264 (9.07)	215 (7.38)	198 (6.80)	167 (5.73)	137 (4.70)	114 (3.91)
1996	315 (10.81)	259 (8.89)	225 (7.72)	210 (7.21)	172 (5.90)	133 (4.57)	126 (4.33)

Cumulative growth rate (% p.a.)

	Portugal	Italy	Germany	Brazil	Spain	United States	France
(a) 1979–96	20.10	3.69	7.39	8.81	6.60	4.12	6.38
(b) 1979–88	42.26	8.01	4.78	(–) 0.45	9.38	3.59	7.59
(c) 1988–96	(–) 0.73	(–) 0.97	10.39	20.26	3.56	4.71	5.04

Note: Figures in brackets represent percentage share of total world import of this product.
Source : As in Table 5.3

in 1996. The dried, salted or smoked form of fish now holds the number one share in the fish imports of Portugal. The reason behind this clear shift in preference needs further investigation. It is observed, however, that during the recent period (1988–96), the import of this product in Portugal is not growing which is also supported by a negative growth rate during the period. It may be that presently Portugal's import of this product is reaching a saturation point after the initial flush in growth during the mid-1980s.

Italy

The share of Italy in the world import of this product has decreased. Italy has also experienced a negative growth rate during the recent period. The share of this product in the total import of fish products in Italy has come down from 21 per cent in 1979 to 10 per cent in 1996, alongside the fall in the share of fresh, frozen or chilled fish in the total import of fish products. There is a clear switch in preference in favour of shellfish and prepared and preserved form of fish products. Italy exhibits the characteristics of a declining market in this fish product.

Germany and France

Germany comes third in respect of the share of the world import of this product, and exhibits a consistent increase in the share of the world import and the rate of growth in import. The share of this product in the total import of fish products in Germany has remained constant at around 9.50 per cent throughout the 18 years. These findings suggest that Germany provides a growing market for this fish product.

France exhibits import characteristics similar to those in Germany, though the share of this product in the total import of fish products is small—at around 4 per cent throughout the period.

Spain

This country's share of the world import of this fish product has increased marginally, but the share of this product in the total import of fish products in Spain has gone down from 14.40 per cent in 1979 to 5.60 per cent in 1996. All the other three fish products have improved their share at the cost of this product. It is likely that Spain's import of this product will further deteriorate.

Brazil

This country stands fourth in terms of share of the world import of this product. But she has increased her share only during the recent two years, otherwise the share can be said to be exhibiting a declining trend. The share of this product in the total import of fish products in Brazil has also declined very sharply, from 78.12 per cent in 1979 to 45.06 per cent in 1996. Brazil has begun exhibiting signs of a declining market for this fish product.

United States

The United States account for a 30 per cent share of American import of this product, but she is showing characteristics of a stagnant market. The share of this product in the total import of fish in the United States has also declined from 2.42 per cent in 1979 to 1.90 in 1996. This country is also a re-exporter of this fish product.

The five countries of Europe discussed here share more than 76 per cent of the European market for this product, while United States and Brazil together control more than 75 per cent of the American market.

Comparative market expansions of this fish product in the countries discussed above are calculated in Table 5.22.

TABLE 5.22
Comparative market (import) expansion of fish in dried, salted or smoked form in major importing countries of America and Europe

(per cent)

Country	Expansion of market	
	1979–96	*1987–96*
Portugal	2150.00	1.29
Italy	85.00	(–) 7.83
Germany	235.82	139.36
Brazil	320.00	275.00
Spain	196.55	59.26
United States	98.51	38.54
France	186.36	51.81
WORLD	128.47	53.64

Asian Countries

Table 5.23 presents data for the import of dried, salted or smoked form of fish for the major Asian nations.

TABLE 5.23
*Import of fish in dried, salted or smoked form
by major importers of Asia*

(in million US dollars)

Year	Japan	Hong Kong	Singapore	Sri Lanka
1979	433 (33.96)	43 (3.37)	17 (1.33)	4 (0.31)
1980	205 (16.53)	50 (4.03)	19 (1.53)	8 (0.65)
1981	275 (20.52)	56 (4.18)	20 (1.49)	3 (0.22)
1982	270 (23.44)	63 (5.47)	22 (1.91)	8 (0.69)
1983	249 (22.41)	60 (5.40)	20 (1.80)	9 (0.81)
1984	212 (21.07)	64 (6.36)	19 (1.89)	15 (1.49)
1985	250 (21.83)	67 (5.85)	16 (1.40)	19 (1.66)
1986	243 (17.29)	83 (5.91)	16 (1.39)	22 (1.57)
1987	325 (17.14)	125 (6.59)	23 (1.21)	26 (1.37)
1988	361 (17.36)	158 (7.60)	30 (1.44)	30 (1.44)
1989	327 (16.41)	149 (7.48)	29 (1.46)	28 (1.40)
1990	363 (15.21)	165 (6.92)	29 (1.21)	33 (1.38)
1991	329 (13.08)	198 (7.87)	39 (1.55)	53 (2.11)
1992	350 (13.74)	271 (10.64)	51 (2.00)	41 (1.61)
1993	380 (16.88)	245 (10.88)	47 (2.09)	40 (1.78)
1994	387 (15.43)	263 (10.49)	47 (1.87)	42 (1.67)
1995	436 (14.97)	340 (11.68)	49 (1.68)	32 (1.10)
1996	444 (15.24)	313 (10.74)	51 (1.75)	33 (1.13)

Cumulative growth rate (% p.a.)

(a) 1979–96	0.15	12.39	6.68	13.22
(b) 1979–88	(–) 2.00	15.56	6.51	25.09
(c) 1988–96	2.62	8.92	6.86	1.20

Note: Figures in brackets represent percentage share of total world import of this product.
Source: As in Table 5.3.

Japan

Japan still holds the major share of the world import of dried, salted or smoked fish, though her share has consistently fallen from a high of 33.96 per cent in 1979 to a low of about 15 per cent in 1996. The rate of growth during the recent period (1988–96) has, however, increased from a negative of 2 per cent during the mid-period (1979-88).

The share of this product in Japan's total import of fish products has also declined from 10.94 per cent in 1979 to 2.66 per cent in 1996. Although Japan will continue to be the major importer of this product from Asia, the country is showing characteristics of a declining market.

Other Countries

The other three countries of Asia listed in Table 5.23 have increased their share of the world import of this product since 1979. The highest share among them is of Hong Kong, whose share has increased from 3.37 per cent in 1979 to 10.74 per cent in 1996. However, though the rate of growth of import of this product in Hong Kong is considerably high, it has decreased during the last period. The share of this product in Hong Kong's total import of fish products has marginally increased from 14.05 in 1979 to 16.34 per cent in 1996. A part of Hong Kong's import of this product is also exported further.

Singapore's share of the world import of this product has increased only marginally, though the compounded rate of growth has remained steady during the entire period. The share of this product in the fish import basket of Singapore has fallen from about 18 per cent in 1979 to 8 per cent in 1996. Singapore is also a re-exporter of this product.

TABLE 5.24

Comparative market (import) expansion of fish in dried, salted or smoked form in major importing countries of Asia

(per cent)

Country	Expansion of market	
	1979–96	*1987–96*
Japan	2.54	36.62
Hong Kong	627.91	150.40
Singapore	200.00	121.74
Sri Lanka	725.00	26.92
WORLD	128.47	53.64

The Sri Lankan market is small in size, but its share has increased consistently. However, there is a considerable fall in the rate of growth during the recent period.

These four countries together control more than 91 per cent of the Asian import of this fish product. The comparative market expansions of this product in the Asian countries are given in Table 5.24.

STRATEGIC MARKETING FOCUS

In Chapter 3, Market Opportunity Indexes for the major fish importing countries of the world have been developed. Here an attempt is made to develop specific models for each of the four groups of fish products to enable an exporting country producing a particular group of fish product to draw up a preferential list of countries in order to have a marketing focus.

As seen earlier, a country may presently have a large share of import market in a particular form of fish product, but her market is no longer expanding or there is a deceleration in expansion of the market. This indicates that either the market is saturating or a distinct switching out is occurring from this particular form of fish product. On the other hand, it has also been observed earlier that in a particular fish product market a country might be presently having a small market share but the expansion rate of the market is very high. Such a country offers a growing market, while the former has a shrinking market. This calls for a strategic refocusing by the exporters of this product. The former type of countries cannot be written off because the market share, though it may be saturating or decelerating, is still substantial. The strategy in such a situation will be the maintenance of the market at its present level of exposure or gradual withdrawal from the market if it is shrinking. An alternative strategy could also be aggressive market penetration in a bid to snatch away a share of the market from other competing countries. The latter strategy is costly and calls for aggressive marketing skills. An exporting country should, therefore, adopt a SWOT approach within a cost–benefit analytical framework before making a choice of a particular strategy.

In the case of the second type of countries who presently have a low market share but whose market is expanding, the strategy should be towards the development of and concentration in the market. Penetration to increase the market share could well fit into this strategy within a framework of SWOT and cost–benefit analysis.

Market Risk

An import market may be expanding with a good share of the world import, but it can be a risky market in terms of sales volatility. An exporter to that market should, therefore, take this volatility factor into consideration while considering strategic focus on an importing country. As seen earlier, high volatility makes it difficult for projecting the operating variables (for example, catch, sales etc.) in a particular market. A business cannot operate without some kind of plan for the future. The major variable on which the plan revolves around is sales. All resources of the business are engaged towards attaining the planned level of sales. If now the sales could not be achieved owing to short supply of inputs (for example, catch) or a sudden dip in sales against the projected level, (for example, fall in the import of fish product by the importing country), then all the costly resources engaged for achieving the planned level of sales (exports) will be unabsorbed, causing severe strain on the bottom line of the business enterprise. It will, therefore, be wrong to decide on a market focus strategy based only on market expansion and market share. Sales or import volatility in the importing country must be factored in any model for market focusing strategy. We have earlier considered measurement of volatility by the coefficient of variation (CV). This factor should now be incorporated in our model.

A market focusing strategy is, therefore, a function of expansion of market in an importing country, market share of the country in world import of a particular fish product, and volatility of imports in the country.

MARKET FOCUS MODEL

The expansion of market in an importing country is measured by developing the Market Expansion Index (MEI) for each country by dividing the percentage expansion of the market in a given period with the total expansion of world market in the same period. For example, in the case of prepared and preserved form of fish product, the world import market has expanded between 1987 and 1996 by 112.09 per cent while the market for this product has expanded by 45.80 per cent in the United States during the same period. Hence, MEI for the United States will be 45.80/112.09, which is equal to 0.41. For all the countries, a period of 10 years (1987–96) is taken for developing the MEI. The market share in the world import of a particular product for a country is taken from the respective table under each product head and incorporated in the model by covering it to decimals. The volatility factor is considered by converting the respective CVs of annual imports (not annual rates of growth) into decimal and then deducting

Table 5.25

Strategic market ranking of importing countries in fresh, frozen or chilled fish product market

Country (1)	Market Expansion Index (1987–96) (2)	World market share (%) (1996) (3)	Market volatility (1–CV) (4)	Comprehensive market parameter (col.2 x 3 x 4) (5)	Market Focus Indicator (%) (6)	Country ranking (7)
United States	0.02	0.1068	0.9433	0.0020	0.28	15
France	0.61	0.0602	0.8631	0.0317	4.53	6
Spain	1.70	0.0588	0.8039	0.0804	11.48	2
Germany	1.17	0.0581	0.7878	0.0535	7.64	3
United Kingdom	0.82	0.0436	0.8106	0.0290	4.14	7
Italy	0.32	0.0436	0.8713	0.0122	1.74	12
Denmark	0.54	0.0258	0.8350	0.0116	1.66	13
Netherlands	1.26	0.0211	0.7830	0.0208	2.97	8
Canada	1.34	0.0163	0.6885	0.0150	2.14	10
Japan	1.32	0.3026	0.7939	0.3171	45.27	1
Korea(R)	2.61	0.0261	0.6786	0.0462	6.60	4
Hong Kong	2.42	0.0230	0.6916	0.0385	5.50	5
Thailand	1.09	0.0217	0.7771	0.0184	2.63	9
Singapore	1.13	0.0104	0.7781	0.0091	1.30	14
Malaysia	2.69	0.0094	0.5889	0.0149	2.12	11
Total		0.8275		0.7004	100.00	

TABLE 5.26

Strategic market ranking of importing countries in prepared and preserved fish product market

Country (1)	Market Expansion Index (1987–96) (2)	World market share (%) (1996) (3)	Market volatility (1–CV) (4)	Comprehensive market parameter (col.2 x 3 x 4) (5)	Market Focus Indicator (%) (6)	Country ranking (7)
United States	0.41	0.1117	0.8983	0.0411	5.76	4
United Kingdom	0.66	0.0884	0.8882	0.0518	7.28	2
France	0.50	0.0804	0.8618	0.0346	4.86	5
Germany	1.03	0.0597	0.7754	0.0477	6.70	3
Italy	0.97	0.0447	0.7888	0.0342	4.80	6
Denmark	0.63	0.0289	0.7973	0.0145	2.03	9
Canada	0.47	0.0242	0.8791	0.0099	1.40	10
Belgium	0.41	0.0237	0.8854	0.0086	1.21	11
Japan	2.70	0.2512	0.6368	0.4319	60.68	1
Hong Kong	1.69	0.0198	0.6877	0.0230	3.23	7
Singapore	2.06	0.0124	0.5922	0.0151	2.12	8
Malaysia	(–) 0.21	0.0030	0.8425	(–) 0.0005	(–) 0.07	12
Total		0.7481		0.7119	100.00	

Table 5.27

Strategic market ranking of importing countries in fresh, frozen or chilled crustaceans, molluscs etc. market

Country (1)	Market Expansion Index (1987–96) (2)	World market share (%) (1996) (3)	Market volatility (1–CV) (4)	Comprehensive market parameter (col.2 x 3 x 4) (5)	Market Focus Indicator (%) (6)	Country ranking (7)
United States	0.39	0.1902	0.8491	0.0630	7.51	4
Spain	1.62	0.0740	0.7913	0.0949	11.31	2
France	0.79	0.0512	0.8614	0.0348	4.15	8
Italy	1.02	0.0461	0.8760	0.0412	4.92	6
Canada	1.50	0.0256	0.7690	0.0295	3.52	9
Denmark	0.43	0.0167	0.8976	0.0064	0.76	15
Belgium	1.58	0.0159	0.7555	0.0190	2.26	11
United Kingdom	0.37	0.0130	0.8393	0.0040	0.48	14
Netherlands	2.76	0.0124	0.6245	0.0214	2.55	10
Germany	1.43	0.0116	0.7973	0.0132	1.57	12
Japan	0.96	0.3843	0.8361	0.3084	36.76	1
Hong Kong	1.40	0.0499	0.8155	0.0570	6.79	5
Korea(R)	13.30	0.0146	0.4759	0.0924	11.02	3
Singapore	1.24	0.0129	0.7754	0.0124	1.48	13
Thailand	18.36	0.0106	0.2124	0.0413	4.92	7
Total		0.9290		0.8389	100.00	

TABLE 5.28

Strategic market ranking of importing countries in dried, salted or smoked fish product market

Country	Market Expansion Index (1987–96)	World market share (%) (1996)	Market volatility (1–CV)	Comprehensive market parameter (col.2 x 3 x 4)	Market Focus Indicator (%)	Country ranking
(1)	(2)	(3)	(4)	(5)	(6)	(7)
Portugal	0.024	0.1081	0.8465	0.0022	0.28	10
Italy	(–) 0.146	0.0889	0.8813	(–) 0.0114	(–) 1.43	11
Germany	2.60	0.0772	0.7327	0.1471	18.47	3
Brazil	5.13	0.0721	0.5220	0.1931	24.24	2
Spain	1.10	0.0590	0.8591	0.0558	7.00	5
United States	0.72	0.0457	0.8892	0.0293	3.68	8
France	0.97	0.0433	0.8590	0.0361	4.53	6
Japan	0.68	0.1524	0.8948	0.0927	11.64	4
Hong Kong	2.80	0.1074	0.7186	0.2161	27.13	1
Singapore	2.27	0.0175	0.7794	0.0310	3.88	7
Sri Lanka	0.51	0.0113	0.7994	0.0046	0.58	9
Total		0.7829		0.7966	100.00	

it from 1. Finally, all these three factors are multiplied to arrive at a comprehensive Market Parameter (MP) for each country as shown in column 5 of Tables 5.25–5.28.

The Market Focus Indicator (MFI) is developed by dividing Market Parameter for each country in a given product market by the total of MPs for all the countries and converting it into percentage form. Each country is then ranked in terms of MFI to indicate strategic focus needed in a particular market for a given fish product.

APPLICATION OF THE MODEL

The Market Focus Indicator (MFI) indicates the attention to be paid by an exporter in a particular importing country of a particular fish product. It also suggests what percentage of additional export should be destinated to a particular country market. For example, the MFI of the United States in fresh, frozen or chilled fish product is only 0.28 per cent, which suggests that additional export of this product to the US market should be 0.28 per cent over and above the present level of exports to this country. To examine why it is so low, we may go back to the import performance of United States in this product. While the world import market in this product has expanded by 98.18 per cent between 1987 and 1996, the US import market has expanded by only 2.03 per cent, despite her having a share of 10.68 per cent of the world import of this product. That is, the US market cannot absorb more than a moderate incremental influx of this product. The US market exhibits low volatility (5.67 per cent), otherwise MFI would have indicated a much lower level of additional export of this product to this market.

As MFIs for all the countries considered here for a particular product are on a 100-point scale, it should be possible for the exporting country (or individual exporter) to develop a portfolio of countries for additional marketing thrusts.

Market Focus Indicators are calculated on the market expansion data for a ten-year period and market share of a country in the terminal year of the period. The MFI model is dynamic over a period of time. As new data for the subsequent year is available, these are incorporated in the table by rejecting the data of the earliest year, and then a fresh set of MFIs is developed.

PROJECTIONS OF WORLD IMPORTS

Fluctuations in the imports of the four forms of fish products have gradually come down during recent years, as is indicated by the coefficient of variations (CV) given at the bottom of Table 5.3. This indicates that markets for these products have stabilized over a period of time. Results of a regression analysis on the import data of all the four products are tabulated in Table 5.29.

TABLE 5.29
Regression coefficients and other parameters for four fish products

Product form	Constant(a)	b–Coefficient	SE(b)	T. Value	R^2*
Fish: fresh, frozen or chilled	1433.28	1209.27	74.87	16.15	0.94
Fish: salted, dried or smoked	774.34	115.43	11.83	9.76	0.86
Shellfish: fresh, frozen or chilled	1897.59	870.59	46.76	18.62	0.96
Fish: prepared, preserved and others	1030.74	432.36	30.01	14.41	0.93

* All significant at 1% level.

On the basis of the parameters of the regression equations shown in Table 5.29, (non-seasonalized) import values of the four forms of fish products are projected in Table 5.30.

TABLE 5.30
Projected levels (non-seasonalized) of world imports of fish products by forms

(in million US dollars)

Projected year	Fish: Fresh, frozen or chilled	Fish: dried or smoked	Shellfish: frozen or chilled	Fish: prepared, preserved and others
2001	29,246	3429	21,921	10,975
2002	30,456	3545	22,792	11,407
2003	31,665	3660	23,663	11,839
2004	32,874	3776	24,534	12,271
2005	34,084	3892	25,404	12,704

We have also run regression analysis on import data of all the four fish products by different countries discussed here. For some countries, which have shown high volatility in their import, results of the regression analysis have not been found satisfactory. Tables 5.31–5.34 give results of the regression analysis for the selected importing countries in different fish products which stood the rigour of statistical testing.

TABLE 5.31
Regression parameters for selected countries of import of fish in fresh, chilled or frozen form

Country	Constant(a)	(b)Coefficient	SE(b)	t. Value	R^2
United States	1150.00	79.88	9.87	8.09	0.80
France	202.38	71.19	7.32	9.73	0.86
Spain	(–) 133.90	85.57	8.12	10.53	0.87
Germany	115.61	68.19	6.37	10.70	0.88
United Kingdom	172.14	47.55	5.42	8.77	0.83
Italy	215.95	51.54	6.92	7.45	0.78
Denmark	44.21	35.27	3.88	9.09	0.84
Netherlands	9.90	26.55	2.78	9.56	0.85
Canada	12.50	17.41	1.53	11.40	0.89
Japan	(–) 305.99	414.59	21.04	19.70	0.96
Korea(R)	(–) 96.77	33.20	2.74	12.11	0.90
Hong Kong	3.70	24.20	2.28	10.61	0.88
Singapore	(–) 1.62	14.72	0.78	18.79	0.96
Malaysia	(–) 27.35	11.63	1.08	10.82	0.88

Note: At 1% level of significance.

TABLE 5.32
Regression parameters for selected countries of import of fish in salted, dried and smoked form

Country	Constant(a)	(b)Coefficient	SE(b)	t. Value	R^2
United States	54.58	4.68	0.40	11.70	0.90
France	25.25	5.22	0.60	8.64	0.82
Spain	31.46	8.69	1.42	6.13	0.72
Germany	16.50	10.01	1.18	8.48	0.82
Hong Kong	(–) 17.82	17.74	1.37	12.95	0.91
Singapore	8.81	2.25	0.27	8.40	0.82
Portugal	5.50	22.48	3.24	6.94	0.75
Sri Lanka	1.66	2.43	0.32	7.57	0.78

Note: At 1% level of significance.

TABLE 5.33
Regression parameters for select countries of import of fish in prepared and preserved form

Country	Constant(a)	(b)Coefficient	SE(b)	t. Value	R^2
United States	264.11	45.47	2.71	16.77	0.95
United Kingdom	200.44	35.08	2.97.	11.80	0.90
France	143.22	33.67	3.33	10.11	0.86
Germany	59.69	26.92	2.74	9.82	0.86
Italy	(–) 21.66	25.34	2.06	12.31	0.90
Denmark	18.75	13.43	1.15	11.67	0.89
Canada	59.02	10.61	0.98	10.81	0.88
Belgium	79.58	8.63	1.22	7.06	0.76
Japan	(–) 379.80	135.93	11.51	11.81	0.90
Hong Kong	(–) 8.03	9.20	0.75	12.26	0.90
Singapore	7.71	4.36	0.69	6.29	0.71

Note: At 1% level of significance.

TABLE 5.34
Regression parameters for select countries of import of crustaceans, molluscs etc. in fresh, frozen or chilled form

Country	Constant(a)	(b)Coefficient	SE(b)	t. Value	R^2
United States	987.05	140.78	8.91	15.79	0.94
France	145.56	41.99	3.32	12.64	0.91
Spain	(–) 116.37	79.27	6.47	12.26	0.90
Germany	3.73	11.89	0.89	13.33	0.92
United Kingdom	43.78	10.93	1.14	9.60	0.85
Italy	15.52	45.24	3.34	13.54	0.92
Denmark	16.93	15.62	1.54	10.13	0.87
Netherlands	(–) 9.24	10.90	1.31	8.33	0.81
Canada	74.27	15.16	1.33	11.36	0.89
Japan	810.87	341.79	24.34	14.05	0.92
Hong Kong	18.67	46.43	2.65	17.51	0.95
Korea(R)	(–) 50.46	12.65	1.59	7.94	0.80
Singapore	(–) 7.47	13.75	0.64	21.52	0.97
Belgium	5.03	15.46	1.42	10.92	0.88

Note: At 1% level of significance.

On the basis of trend equations developed in Tables 5.31–5.34, (non-seasonalized) import of different products by countries are projected in Tables 5.35–5.38.

TABLE 5.35
*Projection (non-seasonalized) of import of fish in fresh,
chilled or frozen form by select countries*

(in million US dollars)

Country	Projection year				
	2001	2002	2003	2004	2005
United States	2987	3067	3147	3227	3307
France	1840	1911	1982	2053	2125
Spain	1834	1920	2005	2091	2177
Germany	1684	1752	1820	1888	1957
United Kingdom	1266	1313	1361	1408	1456
Italy	1401	1453	1504	1556	1607
Denmark	855	891	926	961	997
The Netherlands	621	647	674	700	727
Canada	413	430	448	465	483
Japan	5498	5913	6327	6742	7157
Korea(R)	667	700	733	766	800
Hong Kong	560	585	609	633	657
Singapore	337	352	366	381	396
Malaysia	240	252	263	275	287

TABLE 5.36
*Projection (non-seasonalized) of import of fish in salted,
dried and smoked form by select countries*

(in million US dollars)

Country	Projection year				
	2001	2002	2003	2004	2005
United States	162	167	172	176	181
France	145	151	156	161	166
Spain	231	240	249	258	266
Germany	247	257	267	277	287

(Contd.)

Table 5.36 (contd.)

Country	Projection year				
	2001	*2002*	*2003*	*2004*	*2005*
Hong Kong	390	408	426	443	461
Singapore	61	63	65	67	70
Portugal	523	545	567	590	612
Sri Lanka	58	60	62	65	67

TABLE 5.37

Projection (non-seasonalized) of import of fish in prepared and preserved form by select countries

(in million US dollars)

Country	Projection year				
	2001	*2002*	*2003*	*2004*	*2005*
United States	1310	1355	1401	1446	1492
United Kingdom	1007	1042	1078	1113	1148
France	918	951	985	1019	1052
Germany	679	706	733	759	786
Italy	561	586	612	637	662
Denmark	328	341	355	368	381
Canada	303	314	324	335	346
Belgium	278	287	295	304	313
Japan	2747	2882	3018	3154	3290
Hong Kong	204	213	222	231	240
Singapore	108	112	117	121	126

TABLE 5.38

Projection (non-seasonalized) import of crustaceans, molluscs etc. in fresh, frozen or chilled form

(in million US dollars)

Country	Projection year				
	2001	*2002*	*2003*	*2004*	*2005*
United States	4225	4366	4507	4647	4788
France	1111	1153	1195	1237	1279
Spain	1707	1786	1865	1	2024

(Contd.)

Table 5.38 (contd.)

Country	Projection year				
	2001	*2002*	*2003*	*2004*	*2005*
Germany	277	289	301	313	325
United Kingdom	295	306	317	328	339
Italy	1056	1101	1146	1192	1237
Denmark	376	392	407	423	439
The Netherlands	241	252	263	274	285
Canada	423	438	453	469	484
Japan	8672	9014	9356	9697	10,039
Hong Kong	1086	1133	1179	1226	1272
Korea(R)	241	253	266	278	291
Singapore	309	323	336	350	364
Belgium	361	376	392	407	423

CONCLUSION

Contrary to the existing belief that due to advancement in fish processing technology the value-added, prepared and preserved fish products will push aside the product in fresh, chilled and frozen form and dominate the world market, we find that the expansion of the demand of the latter type of fish product has been the highest during the past two decades. Fish and crustaceans in fresh, chilled or frozen form still constitute more than 77 per cent of world import of total fish products. We also find that another traditional fish product form, namely dried or smoked, continues to make its presence felt in the market. These findings, however, do not belittle the importance of the emergence of the market for high value-added speciality fish products, particularly in America, Europe, and parts of Asia (Japan). What these findings suggest is that the mad rush towards switching over to high value-added, prepared and preserved form, leaving the traditional markets, should be reigned in. This is more true for the developing countries because value addition calls for large capital investment and expenditure in R & D, quality control, and overseas marketing infrastructure, which are not easy to come by in capital scarce economies. It is desirable, therefore, to evolve an optimal marketing strategy for these nations, like India, which will have general focus towards strengthening the marketing of fish in fresh, chilled and frozen form while keeping the speciality fish market in prepared and preserved form for high net worth companies with the necessary managerial and technical skills.

6

World Export of Fish Products and Export Strategy

INTRODUCTION

Despite improvement in surface, water, and air transport systems inter national trade in fish products continues to take place predominantly on an intra-regional basis—a region feeding unto itself. This is more so in the fresh, frozen or chilled form of fish product, and less dominantly so in the prepared and preserved form of fish product. This rule holds as long as a region is self-sufficient in fish production.

In this chapter, first an analysis of the export performance of different fish products in various markets of the world is made. On the basis of these analyses an attempt is then made to develop a methodology for determining the export marketing strategies of fish exporting countries, with particular reference to developing nations.

EXPORT OF FISH IN FRESH, FROZEN OR CHILLED FORM

CONTINENT ANALYSIS

Europe's share of the world import of fish in fresh, frozen or chilled form closely matches with that of the world export, whereas Asia's demand for fish product is so high as compared to its domestic production that its import outmatches the export of this product form by a huge margin. While Europe's net import of this fish product was US$ 583 million, Asia's net import was US$ 6364 million as on 1996. Table 6.1 makes a comparative analysis of the five continents in the international trade of this fish product.

TABLE 6.1

Comparative analysis of international trade of five continents,
in fresh, frozen or chilled form of fish production

Continents	Share of world import (%)		Share of world export (%)		Rise or fall in share of import and export (%)	
	1979	1996	1979	1996	Import	Export
America	26.10	13.85	30.16	22.90	(46.23)	(24.07)
Europe	45.25	36.87	45.80	40.06	(18.52)	(12.53)
Asia	24.43	42.71	19.52	18.24	74.83	(6.56)
Africa	2.71	2.54	2.36	4.76	(6.27)	101.70
Oceania	1.53	0.78	2.71	4.13	(49.02)	52.40

Note: Figures in brackets represent percentage fall in share.

Source: Calculated from *International Trade Statistics Year Book*, Vol. II, *Trade by Commodity*, United Nations, New York, various issues.

America

As discussed earlier and as observed from Table 6.1, America's import of fish in fresh, frozen or chilled form has drastically decreased since 1979. This is primarily due to a change in preference towards prepared and preserved form of fish (or due to an absolute shift in the demand of fish product which we shall examine later). When such a shift occurs initially, this would cause the export of selected items of fish species for which domestic demand has reached the saturation point and the import of those types where domestic demand exists. If domestic demand does not expand then the area becomes a net exporter of the particular fish product. This is precisely what has happened in America. The fish export basket of America contained 37.63 per cent of this product in 1979, and this has increased to 42 per cent in 1996. The product also holds the number one position in the export basket of fish product of America.

Europe

Europe's share of the world import and export in this fish product was almost the same in 1979. However, the share of both import and export has fallen in 1996. Europe is a net importer, but in money terms the net import is comparatively small, which suggests that some countries of Europe are engaged in re-exporting a part of their import of this fish product.

We should point out here that when the net import or net export of a particular product by continent or country bears a small percentage of the total import or export of that place then the continent or the country can be said to be engaged in re-exporting a large part of imports.

The share of this particular fish product in the total exports of fish products from Europe is not only the highest, it has also gone up from 50.75 per cent in 1979 to 55.36 per cent in 1996.

Asia

Asia was a net importer in this product in 1979 by a small amount. But over the past 18 years the net import balance has gone up by a huge amount and also by a sizeable percentage, primarily because of the high expansion of domestic demand which did not allow the share of export to go up during this period. (In fact, there is a 6.5 per cent fall in the share of the world export). This is also supported by the fact that the share of this product in the total fish exports from Asia has also gone down from 30 per cent in 1979 to about 23 per cent in 1996. Asia must be getting a major part of supplies of this fish product from America, followed by Europe and also from the other two continents, though by small amounts.

Africa and Oceania

It may be noticed that exports of this fish product from Africa and Oceania have gone up substantially, as shown in Table 6.1. Both these continents are now net exporters of this fish product.

The share of this fish product in the total exports of fish from Africa has gone up from 22.86 per cent in 1979 to 33.16 per cent in 1996. In the case of Oceania, the percentages are 23.16 and 43.48, respectively.

Opportunity and Threat

The huge net import position of Asia provides both an opportunity and threat to the fish producing countries of Asia. This continent provides a vast market unto itself. The continent must be getting the supplies from high cost producers of America and Europe whereas the cost of production in most Asian countries is comparatively much lower. This competitive advantage can be realized by increasing the fish production in the Asian countries.

One threat, however, is that American and European countries have already established a standard of quality in the Asian market which may increase rejections by the Asian importing countries, particularly Japan, the biggest importer, if exports are made without due regard to the quality requirement. The other threat comes from Africa and Oceania, whose exports are expanding, though during the recent period there is a fall in the exports of Oceania.

COUNTRY ANALYSIS

American and European Countries

Table 6.2 analyses the export performances of the major countries of America and Europe.

Table 6.2 excludes USSR for which consistent data are not available. However, data available for exports of this fish product for the period 1992–6 indicate that her share of export has increased from 6.82 per cent in 1992 to 10.07 per cent in 1996.

Norway

This country has the largest share of the world export of fish in fresh, frozen or chilled form. The share has also increased consistently from 1979. While the world export in this product has expanded by 334 per cent between 1979 and 1996, Norway's export has expanded by about 719 per cent, that is more than double the world expansion. Norway's compounded growth rate is also very impressive, though there has been some fall in the growth rate during the recent period (1988–96).

The share of this product in the export basket of Norway has also gone up substantially, from 40.68 per cent in 1979 to 71.06 per cent in 1996. The share of exports of all other fish products have gone down substantially during the period. It appears that Norway has deliberately changed her export strategy by putting an almost singular emphasis on the export of fish in fresh, frozen or chilled form. This strategy has enabled the nation to become the leader in the export of this fish product.

United States

The United States has increased her share of export in this product during the mid-period, but it has come down in 1996 to the level of 1979. This is also evident in the decline of the cumulative growth rate during 1988–96. However, the United States is able to maintain her average share of export during the entire period of 18 years. It is possible for the United States to do it because the average demand of fish products as a whole is on the decline there. This is indicated by the fact that per capita consumption of fish in the United States has come down to 21.9 kg during 1993–5 from an average of 21.3 kg during 1988–90, while the population of the country has increased by 6.93 per cent between the same periods. This indicates that there is a fall in the expected demand by about 2.75 per cent between 1990 and 1995. Although the figure is small, the one thing that is certain is that the demand of fish in the United States is not growing. It may be that a section of the US consumers, belonging particularly to the younger generation, is getting out

(in million US dollars)

TABLE 6.2

Export of fish in fresh, chilled or frozen form by major exporters of America and Europe

Year	Norway	United States	Denmark	Canada	The Netherlands	Iceland	France	United Kingdom
1979	288 (6.19)	460 (9.89)	469 (10.08)	548 (11.78)	286 (6.15)	319 (6.86)	159 (3.42)	148 (3.18)
1980	274 (5.74)	363 (7.61)	500 (7.61)	553 (11.59)	302 (6.33)	323 (6.77)	200 (4.19)	150 (3.14)
1981	274 (5.44)	545 (10.81)	463 (9.18)	618 (12.26)	319 (6.33)	274 (5.44)	200 (3.97)	125 (2.48)
1982	296 (5.94)	588 (11.79)	455 (9.13)	686 (13.76)	318 (6.38)	264 (5.29)	182 (3.65)	118 (2.37)
1983	346 (7.25)	525 (11.00)	443 (9.28)	619 (12.97)	317 (6.64)	284 (5.95)	191 (4.00)	130 (2.72)
1984	345 (6.96)	520 (10.49)	430 (8.68)	642 (12.95)	322 (6.50)	265 (5.35)	179 (3.61)	121 (2.44)
1985	361 (6.37)	664 (11.71)	457 (8.06)	691 (12.19)	354 (6.24)	331 (5.84)	226 (3.99)	147 (2.59)
1986	537 (7.09)	826 (10.91)	667 (8.81)	915 (12.08)	488 (6.44)	436 (5.76)	292 (3.86)	217 (2.87)
1987	785 (8.33)	1002 (10.63)	853 (9.05)	1047 (11.11)	593 (6.29)	552 (5.86)	381 (4.04)	315 (3.34)
1988	937 (7.46)	1393 (11.10)	832 (6.63)	1135 (9.04)	576 (4.56)	526 (4.19)	431 (3.43)	322 (2.56)
1989	915 (7.58)	1438 (11.90)	761 (6.30)	1041 (8.62)	578 (4.79)	583 (4.83)	480 (3.97)	355 (2.94)
1990	1291 (9.09)	1788 (12.58)	1024 (7.21)	1153 (8.12)	753 (5.30)	762 (5.36)	590 (4.15)	433 (3.04)
1991	1378 (8.42)	1857 (11.34)	1082 (6.61)	1091 (6.67)	812 (4.96)	812 (4.96)	594 (3.63)	510 (3.12)
1992	1461 (9.15)	2124 (13.30)	1056 (6.61)	920 (5.76)	808 (5.06)	746 (4.67)	606 (3.79)	534 (3.34)

(Contd.)

Table 6.2 (contd.)

Year	Norway	United States	Denmark	Canada	The Netherlands	Iceland	France	United Kingdom
1993	1420 (9.41)	1782 (11.81)	1054 (6.99)	875 (5.80)	694 (4.60)	662 (4.39)	493 (3.27)	406 (2.69)
1994	1752 (9.96)	1843 (10.47)	1104 (6.27)	830 (4.72)	695 (3.95)	717 (4.07)	541 (3.07)	463 (2.63)
1995	2067 (10.45)	2078 (10.50)	1053 (5.32)	758 (3.83)	808 (4.08)	692 (3.50)	532 (2.69)	511 (2.58)
1996	2360 (11.69)	1902 (9.42)	965 (4.78)	784 (3.88)	770 (3.81)	671 (3.32)	556 (2.76)	527 (2.61)
Cumulative growth rate (% p.a.)								
(a)1979–96	13.17	8.71	4.34	2.13	6.00	4.47	7.64	7.76
(b)1979–88	14.01	13.10	6.58	8.43	8.09	5.71	11.72	9.02
(c)1988–96	12.24	3.97	1.87	(–) 4.52	3.70	3.09	3.23	6.35

Year	Chile	Argentina	Mexico	Brazil
1979	29 (0.62)	144 (3.10)	18 (0.39)	26 (0.56)
1980	38 (0.80)	136 (2.85)	8 (0.17)	37 (0.78)
1981	47 (0.93)	94 (1.86)	11 (0.22)	43 (0.85)
1982	51 (1.02)	121 (2.42)	22 (0.44)	34 (0.68)
1983	49 (1.03)	81 (1.70)	11 (0.23)	32 (0.67)
1984	37 (0.75)	47 (0.95)	12 (0.24)	24 (0.48)
1985	48 (0.85)	86 (1.52)	16 (0.28)	28 (0.49)

(Contd.)

Table 6.2 (contd.)

Year	Chile	Argentina	Mexico	Brazil
1986	68 (0.90)	144 (1.90)	61 (0.81)	30 (0.40)
1987	110 (1.17)	192 (2.04)	78 (0.83)	31 (0.33)
1988	163 (1.30)	161 (1.28)	84 (0.67)	32 (0.25)
1989	210 (1.74)	165 (1.37)	73 (0.60)	29 (0.24)
1990	324 (2.28)	231 (1.63)	55 (0.39)	27 (0.19)
1991	407 (2.49)	302 (1.84)	73 (0.45)	34 (0.21)
1992	510 (3.19)	280 (1.75)	40 (0.25)	42 (0.26)
1993	502 (3.33)	302 (2.00)	33 (0.22)	58 (0.38)
1994	571 (3.25)	365 (2.07)	29 (0.16)	41 (0.23)
1995	716 (3.62)	502 (2.54)	87 (0.44)	33 (0.17)
1996	736 (3.64)	495 (2.45)	98 (0.49)	31 (0.15)
Cumulative growth rate (% p.a.)				
(a) 1979–96	20.95	7.53	10.48	
(b) 1979–88	21.15	1.25	18.67	
(c) 1988–96	20.74	15.03	1.95	

Note: Figures in brackets represent percentage share of total world import of this fish product.
Source: As in Table 6.1.

of the fish-eating habit in favour of some other protein food, may be meat, or becoming vegetarian. If this trend continues then the United States can be a formidable competitor in the fish market, more so in fresh, chilled or frozen fish products which are less favoured by the upwardly mobile new generation consumers.

This is also evident from the fact that the share of this product in the total exports of fish from the United States has gone up from 42.68 per cent in 1979 to 62.22 per cent in 1996.

It should be recalled here that the United States is still a net importer of this fish product with a margin of US$ 609 million, which constitutes 24.25 per cent of her total import. This indicates that the United States is also engaged in re-exporting this fish product.

Canada

Besides Norway and United States, all the other developed countries of America and Europe, as shown in Table 6.2, have lost their market share. Canada—the other North American country—has, in contrast to the United States, lost her market share substantially between 1979 and 1996—the loss is approximately 67 per cent. Canada's cumulative rate of growth during the recent period (1988–96) has also become negative. At the same time, Canada's share of the world import has not fallen, but has gone up from 1.23 per cent in 1979 to 1.63 per cent in 1996 with a high compounded growth rate of 10.82 per cent per annum during 1979–96.

The above observation is further supported by the fact that the share of this fish product in the total exports of fish products from Canada has gone down from 50 per cent in 1979 to 34.54 per cent in 1996.

All these indicate that perhaps things contrary to those in the United States are happening in Canada inspite of them being neighbouring countries. The main reason behind this might be the increased level of consumption of fish products in Canada as opposed to that in United States.

Denmark, the Netherlands, and Iceland

These three countries are net exporters of fish in fresh, frozen or chilled form. However, their shares of the world export have almost been halved between 1979 and 1996. These three countries have not been able to expand their exports matching with the rate of expansion of the world market in this product (Norway must have made inroads into their export markets). Although their growth rate in exports during 1979–88 was quite impressive, it fell during the recent period (1988–96).

However, this particular fish product still dominates the fish export basket of these three countries, despite its fall in share during the period under study. In the case of Denmark's export of fish product, the share of fish in fresh, frozen or chilled form has gone down from 66.34 per cent in 1979 to 45.45 per cent in 1996. For the Netherlands, the fall is from 58.13 per cent to 55.16 per cent during the same period. For Iceland, the percentage figures are 66.46 and 55.14, respectively.

Except Iceland, the other two countries appear to be engaged in re-exporting a part of their imports of this fish product. Denmark's net export as a percentage of the total export of this product is 37 per cent, while the percentage for the Netherlands is 35.

France and the United Kingdom

These two countries continue to be net importers of this fish product. Their shares of the world export have fallen by about 19 per cent and 18 per cent respectively, which are rather marginal as compared to other European Countries. This indicates a comparative stability in production and export demand of selected items of this fish product.

The share of this product in the total fish exports of the United Kingdom has remained stable at around 47.5 per cent throughout the period. In the case of France, the share has come down marginally from 61 per cent in 1979 to 57 per cent in 1996. The United Kingdom's percentage of net import to total imports is 48 per cent; for France the percentage is about 61 per cent. This indicates that the United Kingdom's level of re-exports is much higher than that of France.

Chile

Among the developing countries of America only Chile has improved her share of export of this product considerably between 1979 and 1996. Presently, the share is 3.64 per cent of the world export. The rate of growth in exports is consistently very high. Chile is also a net exporter of this fish product. The rising trend in export is expected to continue which makes Chile a challenging exporter in the world market of this product.

Chile may be following a deliberate strategy of boosting up her export of fish in fresh, frozen or chilled form and reducing her export in shellfish. It is found that the share of fish in fresh, frozen or chilled form in the total export of fish products from Chile has gone up from about 35 per cent in 1979 to about 70.60 per cent in 1996 while the share of shellfish in fresh, frozen or chilled form (which was highest in 1979) has gone down from 42.68 per cent in 1979 to only 7.77 per cent in 1996.

Argentina, Mexico, and Brazil

Though Argentina was losing her market share during the mid-period, she has picked it up during the later period. However, on the whole Argentina is losing her market share.

Mexico is improving her market share though marginally, while Brazil is losing its market share. The latter is a net importing country while the former is a net exporter of this product.

Table 6.3 makes a comparative analysis of the market expansion by all the countries discussed so far.

TABLE 6.3

Comparative analysis of export market expansion in fresh, chilled or frozen form of fish by countries of America and Europe

(per cent)

Country	Expansion of market	
	1979–96	*1987–96*
Norway	719.44	200.64
United States	313.48	89.82
Denmark	105.76	13.13
Canada	43.07	(–) 25.12
The Netherlands	169.23	29.85
Iceland	110.34	21.56
France	249.69	45.93
United Kingdom	256.08	40.23
Chile	2437.93	569.10
Argentina	243.75	157.81
Mexico	444.44	25.64
Brazil	19.23	0.01
WORLD	333.98	114.23

Asian Countries

Table 6.4 analyses the performances of the Asian countries in the export of fish in fresh, frozen or chilled form.

TABLE 6.4

Export of fish in fresh, chilled or frozen form by major exporters of Asia

Year	China	Korea (R)	Indonesia	Thailand	Japan	Philippines
1979	–	549 (11.80)	10 (0.21)	20 (0.43)	167 (3.59)	46 (0.99)
1980	–	437 (9.16)	20 (0.42)	20 (0.42)	207 (4.34)	75 (1.57)
1981	–	530 (10.51)	21 (0.42)	37 (0.73)	158 (3.13)	55 (1.09)
1982	–	417 (8.36)	30 (0.61)	31 (0.62)	147 (2.95)	26 (0.52)
1983	–	349 (7.32)	20 (0.42)	30 (0.63)	134 (2.81)	34 (0.71)
1984	–	418 (8.43)	15 (0.30)	43 (0.87)	177 (3.57)	24 (0.48)
1985	–	400 (7.05)	19 (0.34)	51 (0.90)	155 (2.73)	23 (0.41)
1986	–	574 (7.58)	24 (0.32)	83 (1.10)	189 (2.50)	24 (0.32)
1987	–	741 (7.86)	45 (0.48)	100 (1.06)	215 (2.28)	29 (0.31)
1988	220 (1.75)	844 (6.72)	84 (0.67)	131 (1.04)	317 (2.53)	34 (0.27)
1989	243 (2.01)	779 (6.45)	111 (0.92)	154 (1.27)	287 (2.38)	54 (0.45)
1990	296 (2.08)	659 (4.64)	177 (1.25)	184 (1.30)	260 (1.83)	36 (0.25)
1991	311 (1.90)	746 (4.56)	254 (1.55)	326 (1.99)	307 (1.88)	32 (0.20)
1992	382 (2.39)	666 (4.17)	282 (1.77)	329 (2.06)	331 (2.07)	29 (0.18)

(Contd.)

Table 6.4 (contd.)

Year	China	Korea (R)	Indonesia	Thailand	Japan	Philippines
1993	425 (2.82)	636 (4.22)	392 (2.60)	336 (2.23)	321 (2.13)	58 (0.38)
1994	661 (3.76)	697 (3.96)	369 (2.10)	349 (1.98)	319 (1.81)	67 (0.38)
1995	907 (4.59)	745 (3.77)	418 (2.11)	410 (2.07)	295 (1.49)	85 (0.43)
1996	867 (4.29)	694 (3.44)	425 (2.11)	403 (2.00)	284 (1.41)	73 (0.36)
Cumulative growth rate (% p.a.)						
(a) 1979–96	–	1.39	24.68	19.32	3.17	2.75
(b) 1979–88	–	4.89	26.68	23.22	7.38	(–) 3.30
(c) 1988–96	18.70	(–) 2.42	22.47	15.08	(–) 1.36	10.02

Year	Singapore	Hong Kong	India	Malaysia
1979	40 (0.86)	37 (0.80)	15 (0.32)	10 (0.21)
1980	62 (1.30)	31 (0.65)	12 (0.25)	9 (0.19)
1981	74 (1.47)	41 (0.81)	19 (0.38)	9 (0.18)
1982	64 (1.28)	35 (0.71)	24 (0.48)	10 (0.20)
1983	77 (1.61)	34 (0.71)	24 (0.50)	9 (0.19)
1984	84 (1.70)	59 (1.19)	19 (0.38)	8 (0.16)
1985	82 (1.45)	80 (1.41)	22 (0.39)	10 (0.18)

(Contd.)

Table 6.4 (contd.)

Year	Singapore	Hong Kong	India	Malaysia
1986	106 (1.40)	89 (1.18)	49 (0.65)	13 (0.18)
1987	142 (1.51)	118 (1.25)	15 (0.16)	16 (0.17)
1988	189 (1.51)	162 (1.29)	25 (0.20)	19 (0.15)
1989	180 (1.49)	195 (1.61)	33 (0.27)	22 (0.18)
1990	217 (1.53)	194 (1.37)	53 (0.37)	33 (0.23)
1991	230 (1.41)	181 (1.11)	64 (0.39)	35 (0.21)
1992	225 (1.41)	186 (1.16)	93 (0.58)	52 (0.33)
1993	225 (1.49)	174 (1.15)	103 (0.68)	63 (0.42)
1994	253 (1.44)	231 (1.31)	173 (0.98)	56 (0.32)
1995	291 (1.47)	179 (0.90)	196 (1.00)	55 (0.28)
1996	280 (1.39)	137 (0.68)	135 (0.67)	60 (0.30)

Cumulative growth rate (% p.a.)

(a) 1979–96	12.13	8.00	13.80	11.12
(b) 1979–88	18.83	17.83	5.84	7.39
(c) 1988–96	5.04	(–) 2.07	23.47	15.46

Note: Figures in brackets represent percentage to total world export in this product.
Source: As in Table 6.1.

China

Among the countries of Asia, China has the largest share of the world export in fresh, frozen or chilled form of fish product. Her share of the world export has gone up from 1.75 per cent in 1988 to 4.29 per cent in 1996—a rise of about 145 per cent. China's cumulative growth has also been formidable—18.70 per cent during the recent years. But, as mentioned earlier, China's rate of domestic consumption of fish products is on a steep rise due to a rise in income, which might turn the country into a net importer in future. The pressure is, perhaps, the highest in the crustaceans products as the income level of Chinese people is on the rise. The share of this product in the total export of fish products from China has gone down drastically from 72.55 per cent in 1988 to only 27.40 per cent in 1996. In order to maintain her position in the export market, China has adopted an alternative strategy of increasing her export of fish in fresh, frozen or chilled form. This is evidenced by the fact that the share of this product in the fish export basket of China has gone up from 22.70 per cent in 1988 to 30.36 per cent in 1996.

Korea (R)

Korea (R) is the second largest exporter of this fish product from Asia. But her share is falling during the last 18 years—the fall is by about 71 per cent. The rate of growth has also become negative during the recent period (1988–96).

The share of this product in the total export of fish products from the country has also fallen drastically, from 69 per cent in 1979 to 46.44 per cent in 1996.

Korea (R) is also one of the major Asian importers of this fish product. Korea (R)'s export and import are getting closely balanced, which suggests, *inter alia*, that the country might have emerged as a re-exporter of this product.

Indonesia and Thailand

These two countries have considerably increased their share of the world export during the past 18 years. Their rates of growth in the export of this commodity are also among the highest. Thailand is still a net importer, though she is improving her balance of trade gradually, which may be indicating a re-exporting status for the country. Indonesia has, however, turned out to be a net exporter by a sufficiently good margin, though she was a net importer in the early part of the 1980s.

The share of this product in the total fish exports of Indonesia has gone up considerably from 4.53 per cent in 1979 to 25.33 per cent in 1996. In the case of Thailand, the increase is rather marginal—the share has gone up from 6.40 per cent in 1979 to 9.20 per cent in 1996.

Japan

Japan has the largest share in the world import of this fish product; the share is also increasing over time. Japan's share of export of this product is small; it is also shrinking since 1979, when she was the second largest exporter of this product from Asia. The fall in the market share is by about 61 per cent during the period under consideration. Japan has also exhibited a negative growth rate of exports during the recent period.

It is likely that Japan may be exporting some special type of fish in fresh, frozen or chilled form (or is engaged in re-export). There is some evidence to support this conclusion in the fact that the share of this fish product in the total fish exports from Japan has increased from 27.50 per cent in 1979 to 41.30 per cent in 1996.

Philippines

Philippines' size of the export of this fish product is low; her share of the world export is also falling, but the volumes of export and import in this fish product are coming closer during the recent period, which suggests that the country is a re-exporter of this product.

Singapore

Singapore's share of export is increasing since 1979. The increase is by about 62 per cent between 1979 and 1996.

The share of this product in the total export of fish products from Singapore has gone up from 40.82 per cent in 1979 to 51.85 per cent in 1996. This is also the largest item in the fish export basket of Singapore.

Singapore's import performance and import volume in this product are closely matching with that of export, which suggests a re-exporter status of the country.

Hong Kong

Hong Kong's share of the world export of this product has increased consistently from 1984 to 1994. But during the last two years she has lost a good part of her market share. The growth rate in exports has also become negative during the recent period (1988–96), from a high of 17.83 per cent during the mid-period.

Similar to the case of Singapore, in Hong Kong also the share of this product in the total export of fish products has gone up from 19 per cent to 28.42 per cent in 1996.

Hong Kong is, however, a net importer of this product. The export–import behaviour of Hong Kong suggests that she is also engaged in re-exporting the product.

India and Malaysia

These two countries' export performances in this product are not very encouraging. While Malaysia is a net importer by a substantial margin, India's high level of domestic consumption of this product may be acting as hindrance to increase her share of export to which we shall come later.

India does not also appear to be paying much attention to developing this fish product as an exportable commodity. Her export is predominantly shrimp oriented in as much as that shrimp constitutes about 66 per cent of the export basket of India while fish in fresh, frozen or chilled form accounts for only about 15 per cent of the export basket in value terms.

It may be noticed that those countries of Asia who are emerging as re-exporters of this fish product belong to the advanced developing economies of the world. They have better freezing facilities and quality standards acceptable to sophisticated developed countries. Other developing nations of the Asian region may be routing their products, knowingly or unknowingly, through these countries.

TABLE 6.5

Comparative export market expansion in fresh, frozen or chilled form of fish in select countries of Asia

(per cent)

Country	Expansion of market	
	1979–96	1987–96
China	–	294.09
Korea (R)	26.41	(–) 6.34
Indonesia	4150.00	844.44
Thailand	1915.00	303.00
Japan	70.06	32.09
Philippines	58.70	150.71
Singapore	600.00	97.18
Hong Kong	270.27	16.10
India	800.00	800.00
Malaysia	500.00	275.00
WORLD	333.98	114.23

Table 6.5 gives the comparative market expansion of this product by all the countries discussed above.

MARKET COMPETITIVE INDEX

In Chapter 7, indexes of importing countries have been developed for purpose of enabling an exporting country to draw up a list of countries under each product-group for preferential marketing focus. An exporter (country) will be able to realize the opportunities available in an importing country in terms of its own competitive advantages.

Similar to the opportunities provided by an importing country, the market competitiveness of an exporting country also depends upon the rate of her market expansion, market share, and volatility in her exports.

The rate of market expansion is measured as an index to world expansion for last ten years. In the present case it is between 1987 and 1996. When the following year's data is available it may be included in the model after rejecting the earliest year's data to maintain the dynamism of the model.

The market Expansion Index (MEI) is modified by the Market Share (MS) of world export of the latest year for which data is available. This is similarly replaced when the following year's data is available.

An exporting country is constrained by the volatility of her exports which may stem either from domestic catch volatility or production volatility or the import volatility of the countries to which the exports are presently consigned. Volatility is measured by the Coefficient of Variation (CV) and factored in the model as $(1 - CV)$.

The Comprehensive Market Parameter (CMP) is developed as a multiplicative function of the above three variables. That is,

$$CMP = MEI \quad x \quad MS \quad x \quad (1 - CV)$$

for each country in all the four groups of fish products. Finally, Market Competitive Indexes (MCIs) are developed for all the countries discussed under each product-group by dividing the CMP of individual countries with the summation of all CMPs and then multiplying the resultant figure by 100. All the countries are then ranked in the descending order of their respective MCIs. Table 6.6 calculates the MCIs of countries engaged in the export of fish in fresh, frozen or chilled form.

EXPORT OF FISH IN PREPARED OR PRESERVED FORM

While Asia remains the largest net importer of fish in fresh, chilled or frozen form, it is the largest net exporter of prepared or preserved form of fish products. Asia, after satisfying the demand from this region, mainly coming

TABLE 6.6

Market Competitive Index and ranking of major countries engaged in the export of fish in fresh, frozen or chilled form

Country	Market Expansion Index (1987–96)	World market share (%) (1996)	Market volatility (1 − CV)	Comprehensive market parameter (col. 2 x 3 x 4)	Market Competitive Index	Country ranking
(1)	(2)	(3)	(4)	(5)	(6)	(7)
Norway	1.76	0.1169	0.6996	0.1439	24.85	1
United States	0.79	0.0942	0.8698	0.0647	11.17	4
Denmark	0.11	0.0478	0.8571	0.0047	0.81	14
Canada	(−) 0.22	0.0388	0.8482	(−) 0.0072	(−) 1.24	21
Netherlands	0.26	0.0381	0.8780	0.0087	1.50	11
Icelands	0.19	0.0332	0.8777	0.0055	0.95	13
France	0.40	0.0276	0.8966	0.0099	1.71	10
United Kingdom	0.35	0.0261	0.8380	0.0077	1.33	12
Chile	4.98	0.0364	0.5834	0.1058	18.27	2
Argentina	1.38	0.0245	0.6222	0.0210	3.63	7

(Contd.)

Table 6.6 (contd.)

Country (1)	Market Expansion Index (1987–96) (2)	World market share (%) (1996) (3)	Market volatility (1 – CV) (4)	Comprehensive market parameter (col. 2 x 3 x 4) (5)	Market Competitive Index (6)	Country ranking (7)
Mexico	0.22	0.0049	0.6259	0.0007	0.12	19
Brazil	0.0001	0.0015	0.7525	0.0001	0.02	20
China	2.57	0.0429	0.4777	0.0527	9.10	5
Korea (R)	(−) 0.06	0.0344	0.9132	(−) 0.0019	(−) 0.33	22
Indonesia	7.39	0.0211	0.5551	0.0866	14.96	3
Thailand	2.65	0.0200	0.6556	0.0347	6.00	6
Japan	0.28	0.0141	0.9294	0.0037	0.64	16
Philippines	1.32	0.0036	0.6315	0.0030	0.52	17
Singapore	0.85	0.0139	0.8487	0.0100	1.73	9
Hong Kong	0.14	0.0068	0.8676	0.0008	0.14	18
India	7.00	0.0067	0.4123	0.0193	3.33	8
Malaysia	2.41	0.0030	0.6392	0.0046	0.79	15
TOTAL	–	0.6565	–	0.5790	100.00	

from Japan, has enough left to satisfy a large part of the net import demand of this product from America and Europe. This product is technology intensive, requires a higher level of capital investment, and is highest in value addition among all the four groups of fish products. As mentioned before, this product is aimed at the upper-end of the market, which generally is obtainable in the developed countries of the world. Besides Japan (which is also the largest importer of this product), there are not many countries in Asia which can absorb this highly priced product range. It is natural, therefore, that in Asia these products are produced mainly for export markets by advanced developing nations who can afford the high cost of the processing technology and skilled manpower. Similar is the case with Africa which is also a net exporter of this fish product. All the other continents are net importers.

During the later part of the 1970s and the early part of the 1980s, America was a net exporter of the prepared and preserved fish product. But with the rise in the domestic demand of this product and expansion of world market, America turned out to be a net importing continent. The share of world export and import markets in America of this product during recent times remain more or less the same, though their respective shares have fallen between 1979 and 1996, as will be seen later.

TABLE 6.7

Comparative analysis of five continents in the international trade of prepared and preserved fish products

Continents	Share of world import (%)		Share of world export (%)		Rise or fall in share of import and export (%)	
	1979	1996	1979	1996	Import	Export
America	19.03	16.28	27.70	14.68	(14.45)	(47.00)
Europe	51.11	43.45	37.07	28.03	(14.99)	(24.39)
Asia	14.37	31.93	26.36	43.85	122.20	66.35
Africa	10.09	1.39	7.40	7.45	(86.22)	0.68
Oceania	5.44	2.45	1.47	1.65	(54.96)	12.24

Note: Figures in brackets represent percentage fall.
Source: As in Table 6.1.

Europe continues to be a net importer of this fish product. We have seen earlier that this continent is the largest importer, having a share of more than 43 per cent of the world import. In exports, Europe holds the second largest share of the world market—about 28 per cent as on 1996, which, however, has decreased from 37 per cent in 1979.

Oceania is a net importer all through the period. But, while the import share of this continent has fallen by more than half between 1979 and 1996, the share of export has increased, though marginally. Table 6.7 makes a comparative analysis of five continents in the international trade of this product.

CONTINENT ANALYSIS

America

It may be noticed that America's fall in the share of the world export is more than three times that of the import. This suggests that the continent is forced to yield the market place to other competitors. While the world export market in this product has expanded by 332.57 per cent between 1979 and 1996, America's export has grown by only 129.19 per cent during the period.

The share of prepared and preserved fish product in the total export of America has also come down from 16.17 per cent in 1979 to 12.50 per cent in 1996.

If we look at the volume of export and import of this product by America we get an indication that America is engaged in re-exporting a part of imports. For example, in 1996 America's volumes of import and export are US$ 1568 million and US$ 1382 million, respectively. However, this needs further investigation as a country may import a particular range of product and export a different range both belonging to the same broad group.

Europe

Europe's fall in the share of import is more or less the same with that of America. Both continents are experiencing a fall in the share of export, but Europe's fall is not as extreme as that of America.

The share of this fish product in the fish export basket of Europe has gone down only marginally from 19.22 per cent in 1979 to 18.07 per cent in 1996.

The processed fish market has a number of differentiated product ranges. A country/continent could be importing a particular range of speciality products and at the same time exporting a different range of products belonging to the same group. As Europe has a large number of small nation states who have different tastes for speciality fish products, both imports and exports in sizeable quantities do take place to cater to the varying tastes of the region. In 1996, Europe's export and import of this product-group was US$ 2640 million and US$ 4184 million, respectively.

Asia

Asia's shares of both import and export of this fish product have gone up between 1979 and 1996. The rise in import share is, however, nearly double that of export. However, Asia has expanded its export market by 619.34 per cent between 1979 and 1996 as compared to 332.57 per cent in the world export of this product. The share of this product in the total fish exports from Asia has also gone up from 18.94 per cent in 1979 to 26.08 per cent in 1996.

This continent is emerging as a formidable leader in the export market of this speciality fish product. This has been possible due to the liberal practices of some of the advanced developing nations of the region in international trade.

Africa and Oceania

Africa, in a small way, is a new competitor in this product market. The fall in the share of import of this product is quite substantial, while there is a small rise in the share of export. Africa's share of export has expanded by 336.02 per cent which more than matches the expansion of the world export in this product. However, Africa might be changing her marketing strategy and concentrating more on the export of crustaceans etc. in fresh, frozen or chilled form. The share of fish in prepared and preserved form in the total fish exports from Africa has fallen from 33.47 per cent in 1979 to 24.20 per cent in 1996, while that of crustaceans etc. has gone up from 38.25 per cent in 1979 to 41.12 per cent in 1996.

Although Oceania's share of export is much smaller than that of Africa, the continent's percentage rise in the share of world export is much more than that of Africa. Oceania's expansion of export between 1979 and 1996 is 384.38 per cent—more than the world expansion. The share of this product in the fish export basket of Oceania is being consistently maintained at around 8 per cent throughout the period.

COUNTRY ANALYSIS

American and European Countries

Table 6.8 analyses the export performance of the major exporting countries of America and Europe.

From Table 6.8 it is seen that the European countries are the leaders in the export of prepared and preserved fish products among the two

Table 6.8
Export of preserved and prepared fish product by major exporting countries of America and Europe

(in million US dollars)

Year	Denmark	The Netherlands	Spain	United States	Canada	Norway
1979	126 (5.79)	78 (3.58)	127 (5.83)	142 (6.52)	282 (12.95)	133 (6.11)
1980	143 (5.44)	80 (3.04)	136 (5.17)	208 (7.91)	185 (7.04)	168 (6.39)
1981	144 (5.54)	70 (2.69)	170 (6.54)	72 (2.77)	251 (9.65)	154 (5.92)
1982	158 (6.75)	74 (3.16)	93 (3.97)	43 (1.84)	200 (8.54)	148 (6.32)
1983	166 (6.70)	80 (3.23)	71 (2.86)	40 (1.61)	245 (9.88)	204 (8.23)
1984	160 (6.47)	76 (3.07)	72 (2.91)	35 (1.42)	221 (8.94)	185 (7.48)
1985	180 (6.33)	86 (3.02)	77 (2.71)	112 (3.94)	237 (8.33)	184 (6.47)
1986	282 (7.25)	128 (3.29)	77 (1.98)	141 (3.63)	299 (7.69)	221 (5.69)
1987	323 (7.43)	172 (3.95)	101 (2.32)	136 (3.13)	417 (9.59)	231 (5.31)
1988	364 (6.11)	173 (2.90)	108 (1.81)	192 (3.22)	249 (4.58)	188 (3.16)
1989	355 (6.73)	195 (3.70)	121 (2.29)	306 (5.80)	248 (4.70)	219 (4.15)
1990	460 (7.95)	270 (4.67)	126 (2.18)	304 (5.25)	222 (3.84)	239 (4.13)
1991	464 (7.29)	274 (4.30)	145 (2.28)	370 (5.81)	244 (3.83)	218 (3.42)
1992	488 (7.67)	298 (4.68)	146 (2.30)	398 (6.26)	234 (3.68)	239 (3.76)

(Contd.)

Table 6.8 (contd.)

Year	Denmark	Netherlands	Spain	United States	Canada	Norway
1993	470 (7.13)	296 (4.49)	153 (2.32)	351 (5.33)	210 (3.19)	198 (3.00)
1994	483 (6.37)	322 (4.25)	186 (2.46)	332 (4.38)	223 (2.94)	211 (2.78)
1995	511 (5.82)	369 (4.21)	262 (2.99)	325 (3.71)	225 (2.56)	204 (2.33)
1996	483 (5.13)	333 (3.54)	293 (3.11)	293 (3.11)	266 (2.82)	214 (2.27)
Cumulative growth rate (% p.a.)						
(a) 1979–96	8.23	8.91	5.04	4.35	(–) 0.34	2.84
(b) 1979–88	12.51	9.25	(–) 1.78	3.41	(–) 1.37	3.92
(c) 1988–96	3.60	8.53	13.29	5.43	0.83	1.63

Year	Iceland	Chile	France	Ecuador	United Kingdom	Greenland	Portugal
1979	9 (0.41)	14 (0.64)	32 (1.47)	27 (1.24)	44 (21.02)	41 (1.88)	77 (3.54)
1980	10 (0.38)	24 (0.91)	36 (1.37)	74 (2.81)	55 (2.09)	60 (2.28)	97 (3.69)
1981	9 (0.35)	25 (0.96)	34 (1.31)	53 (2.04)	50 (1.92)	53 (2.04)	70 (2.69)
1982	13 (0.56)	25 (1.07)	32 (1.37)	47 (2.01)	48 (2.05)	34 (1.45)	70 (2.99)
1983	15 (0.61)	32 (1.29)	33 (1.33)	28 (1.13)	56 (2.26)	32 (1.29)	69 (2.78)
1984	16 (0.64)	40 (1.62)	37 (1.50)	16 (0.65)	53 (2.14)	38 (1.54)	64 (2.59)

(Contd.)

Table 6.8 (contd.)

Year	Iceland	Chile	France	Ecuador	United Kingdom	Greenland	Portugal
1985	17 (0.60)	41 (1.44)	39 (1.37)	180 (6.32)	55 (1.93)	37 (1.30)	62 (2.18)
1986	21 (0.54)	75 (1.93)	57 (1.47)	308 (7.92)	68 (1.75)	72 (1.85)	73 (1.88)
1987	29 (0.67)	110 (2.53)	76 (1.75)	36 (0.83)	80 (1.84)	84 (1.93)	75 (1.72)
1988	31 (0.52)	109 (1.83)	80 (1.34)	27 (0.46)	85 (1.43)	83 (1.39)	80 (1.34)
1989	26 (0.49)	113 (2.14)	81 (1.54)	27 (0.51)	81 (1.54)	86 (1.63)	97 (1.84)
1990	29 (0.50)	117 (2.02)	92 (1.59)	31 (0.54)	96 (1.66)	118 (2.04)	116 (2.00)
1991	23 (0.36)	139 (2.18)	91 (1.43)	29 (0.46)	104 (1.63)	91 (1.43)	119 (1.87)
1992	24 (0.38)	147 (2.31)	94 (1.48)	44 (0.69)	102 (1.60)	105 (1.65)	110 (1.73)
1993	115 (1.75)	174 (2.64)	86 (1.31)	72 (1.09)	119 (1.81)	103 (1.56)	90 (1.37)
1994	169 (2.23)	147 (1.94)	105 (1.39)	103 (1.36)	142 (1.87)	93 (1.23)	113 (1.49)
1995	223 (2.54)	183 (2.09)	152 (1.73)	118 (1.35)	137 (1.56)	119 (1.36)	137 (1.56)
1996	204 (2.17)	182 (1.93)	171 (1.82)	151 (1.60)	140 (1.49)	115 (1.22)	108 (1.15)
Cumulative growth rate (% p.a.)							
(a) 1979–96	20.15	16.29	10.36	10.66	7.05	6.25	2.01
(b) 1979–88	14.73	25.61	10.72	0.00	7.59	8.15	0.43
(c) 1988–96	26.56	6.62	9.96	24.01	6.44	4.16	3.82

Note: Figures in brackets represents percentage share of total world export in this fish product.
Source: As in Table 6.1.

continents. The United States and Canada are ranked in the fourth and fifth positions only.

The table excludes USSR due to a lack of consistency in the data. However, the recent data of the Russian Federation indicate that the Federation is a net importer of this fish product. Her import stands at US$ 175 million and export at US$ 88 million as on 1996.

Table 6.8 also indicates that the real spurt in the export of this product has occurred from 1987–8 in the majority of the countries. As for the world's exports, it has gone up by 116.48 per cent in ten years' time, while the total expansion of export between 1979 and 1996 is 332.57 per cent.

Denmark

The country has the largest share of the world export among the American and European countries in this fish product. The share of this product in the total exports of fish products from Denmark has also gone up from 17.82 per cent in 1979 to 22.76 per cent in 1996. But her share of the world export of this product has fallen marginally in 1996 from what she had in 1979. Denmark's share went up to as high as 7.95 per cent in 1990, but from 1994 the share began to decline. Between 1979 and 1996, Denmark's export has expanded by about 283 per cent which is lower than the expansion in world export by 50 percentage points. The compounded growth rate in Denmark's export has gone down substantially during the recent period (1988–96) as compared to the mid-period. we may recall that Denmark's import of this product has also started growing from 1986 and touched a level of US$ 278 million in 1996, recording an expansion of 126 per cent between 1986 and 1996. It is likely that Denmark is experiencing an expansion of domestic demand in this product, or it is concentrating more on expanding the export in crustaceans etc., whose share in the total export of fish products has gone up from 8.50 per cent in 1979 to 18.20 per cent in 1996. The market share of Denmark is expected to slide down further in the years to come. It is a signal for other exporting countries to plan for taking over those markets which are hitherto being served by Denmark.

The Netherlands

Netherland's case is not similar to that of Denmark, though in 1996 her market share came down to the market share in 1979. There was a spurt in the growth of the market share from 1987, which continued till 1995. But in 1996 the share fell to the 1979 level. The general trend suggests that the Netherlands will retrace her position and be on the expansion path once again. The share of this product in the fish export basket of the Netherlands has also gone up from 15.85 per cent in 1979 to 23.85 per cent in 1996. The country is maintaining a more or less stable rate of growth of around 9 per cent per annum. The trend is likely to continue. The Netherland's import of this group of fish product is growing at a steady rate of around 5 per cent per annum, which is half that of Denmark's rate. The Netherlands is expected to take over Denmark's number one position in the near future.

United States and Canada

Except Iceland, Chile, France, and Ecuador, all the other countries of America and Europe reported in Table 6.8 have lost their share of the world export. Canada's loss of the market share is the largest. The United States is another big loser in America. Between 1979 and 1996 United States has lost 52.30 per cent and Canada 78.22 per cent of their share of the world export in prepared and preserved fish products.

The United States is also a net importer of this product by a sizeable amount of US$ 783 million as on 1996, which suggests an expansion of the demand of this product in United States. As compared to this, Canada's net import position is only US$ 33 million. The gap between export and import is closing up in Canada which indicates that Canada might be importing this product for re-export to the United States. However, after the initial expansion of demand, the growth of imports in the United States is showing a sharp declining trend during the recent period (1988–96). As a result, Canada's export is also declining at a fast rate. The cumulative rate of growth of Canada's export for the entire period (1979–96) is negative, which, however, has improved to a small positive rate during the recent period. In both the two countries, the share of this product in their total export of fish products has declined. For the United States, the fall is from 13.17 per cent in 1979 to 9.58 per cent in 1996. In the case of Canada the figures are 25.50 per cent and 11.72 per cent for the respective years. All these findings suggest that both the United States and Canada are losing out in the competition in the export market of prepared and preserved fish product.

Norway and Portugal

In Europe, the highest loss of market share is that of Portugal (67.51 per cent), followed by that of Norway (62.85 per cent)—the latter's share of the world export is, however, double that of the former.

Norway's import of this product is small—only US$ 54 million as of 1996. The country is a net exporter. In 1983, Norway was the second largest exporter of this product in the whole of America and Europe, but she began losing the market share since then. The share of this product in the fish export basket of Norway has also fallen from 18.80 per cent in 1979 to 6.45 per cent in 1996. These trends suggest that she will be losing further. The rate of growth in exports during the recent period has also come down drastically.

Portugal's case is similar to that of Norway. She is a net exporter with a low level of import—only US$ 28 million as on 1996. Her share of export has been falling consistently since 1980, though the compounded rate of growth has improved during the recent period (1988–96) but is far below the world expansion rate of the market. It is unlikely that Portugal will regain her position in the export market in future.

United Kingdom and Greenland

These two countries have lost their share of the world export, though the loss is not as high as that of the countries discussed above. The rates of growth in exports for the entire period (1979–96) for the two countries are quite good. A fall is, however, noticed during the recent period (1988–96).

The United Kingdom is a net importer of this product by a sizeable margin. Import has expanded by about 258 per cent—closest to the world expansion rate in this product. This indicates that the domestic demand of this product is on the rise in the United Kingdom, which leaves a lesser amount for export. The export of this product, which accounted for 14.15 per cent of the total export of fish products from the United Kingdom, has declined to 12.60 per cent in 1996.

Greenland is a net exporter. Her import of this fish product is negligible. Greenland suffers from seasonal fluctuations in her export which may be due to her fluctuating catches, but the overall trend suggests that she will recover her position and move ahead.

Other Countries

The quantum and share of the world export of Iceland, Chile, France, and Ecuador are small but all of them exhibit a growth in their respective shares. Their cumulative growth rates are also quite substantial, except that Ecuador suffered from a zero growth in the mid-period but recovered considerably during the recent period (1988–96). France is a net importing country with a substantial margin while the remaining three countries are net exporters, with a rising share in the world trade. In all these countries the share of this product in their export basket of fish products have gone up substantially between 1979 and 1996. All these trends suggest that these four countries will increase their share of exports further.

Table 6.9 makes a comparative analysis of the market expansion of the exporting countries from America and Europe.

TABLE 6.9
Comparative analysis of market expansion by select countries of America and Europe in prepared and preserved fish products

(per cent)

Country	Expansion of market	
	1979–96	1987–96
Denmark	283.33	49.54
The Netherlands	326.92	93.60
Spain	130.71	190.10
United States	106.34	115.44
Canada	(–) 5.67	(–) 36.21
Norway	60.90	(–) 7.36
Iceland	2166.67	603.45
Chile	1200.00	65.45
France	434.38	125.00
Ecuador	459.26	319.44
United Kingdom	218.18	75.00
Greenland	180.49	36.90
Portugal	40.26	44.00
WORLD	332.57	116.48

Asian and African Countries

Table 6.10 analyses the export performance of the major exporting nations of Asia and Africa.

Thailand

Thailand is the largest supplier of prepared and preserved fish products to the world. Presently, the country accounts for about 17 per cent of the world export of this product, which has increased from about 3 per cent in 1979—an expansion in share by about 440 per cent during the period. The cumulative growth rate in the export of this product is a formidable 20.36 per annum for the whole period, though during the recent period (1988–96) there is a decline in the growth rate. The share of this product in Thailand's total export of fish products has also gone up considerably—from 21.40 per cent in 1979 to 35.17 per cent in 1996. Thailand's high product quality, high productivity of labour, and excellent production facilities have contributed towards the high growth rate in processed fish items. Some of these factors discussed earlier are elaborated further in the following paragraphs.

TABLE 6.10

Export of prepared and preserved fish product by major exporting countries of Asia and Africa

(in million US dollars)

Year	Thailand	China	Korea (R)	Japan	Morocco
1979	67 (3.08)	–	54 (2.48)	376 (17.27)	77 (3.54)
1980	87 (3.31)	–	53 (2.02)	618 (23.50)	85 (3.23)
1981	108 (4.15)	–	69 (2.65)	519 (19.95)	90 (3.46)
1982	150 (6.41)	–	82 (3.50)	455 (19.44)	66 (2.82)
1983	187 (7.54)	–	91 (3.67)	406 (16.38)	55 (2.22)
1984	264 (10.68)	–	115 (4.65)	438 (17.72)	63 (2.55)
1985	285 (10.02)	–	136 (4.78)	426 (14.97)	72 (2.53)
1986	445 (11.45)	–	228 (5.87)	439 (11.29)	96 (2.47)
1987	555 (12.76)	–	265 (6.09)	377 (8.67)	94 (2.16)
1988	828 (13.90)	36 (0.60)	366 (6.14)	354 (5.94)	110 (1.85)
1989	848 (16.08)	47 (0.89)	303 (5.74)	340 (6.45)	118 (2.24)
1990	993 (17.16)	54 (0.93)	305 (5.27)	309 (5.34)	142 (2.45)
1991	1188 (18.68)	85 (1.34)	306 (4.81)	316 (4.97)	155 (2.44)
1992	1151 (17.47)	202 (3.07)	294 (4.46)	313 (4.75)	159 (2.41)
1993	1205 (15.90)	296 (3.91)	344 (4.54)	303 (4.00)	146 (1.93)
1994	1489 (19.65)	523 (6.90)	362 (4.78)	285 (3.76)	158 (2.08)
1995	1590 (16.97)	788 (8.98)	395 (4.50)	277 (3.15)	181 (2.06)
1996	1565 (16.61)	1131 (12.01)	417 (4.43)	261 (2.77)	176 (1.87)

Cumulative growth rate (% p.a.)

(a) 1979–96	20.36	–	12.78	(–) 2.12	4.98
(b) 1979–88	32.23	–	23.69	(–) 0.67	4.04
(c) 1988–96	8.28	53.87	1.64	(–) 3.74	6.05

(Contd.)

Table 6.10 (contd.)

Year	Philippines	Malaysia	Indonesia
1979	8 (0.37)	17 (0.78)	1 (0.05)
1980	31 (1.18)	46 (1.75)	1 (0.04)
1981	54 (2.08)	65 (2.50)	4 (0.15)
1982	49 (2.09)	51 (2.18)	2 (0.08)
1983	62 (2.50)	57 (2.30)	5 (0.20)
1984	47 (1.90)	45 (1.82)	5 (0.20)
1985	49 (1.72)	46 (1.62)	3 (0.10)
1986	53 (1.36)	57 (1.47)	5 (0.13)
1987	60 (1.38)	72 (1.66)	12 (0.28)
1988	100 (1.68)	72 (1.21)	27 (0.45)
1989	115 (2.18)	81 (1.54)	49 (0.93)
1990	102 (1.76)	75 (1.30)	55 (0.95)
1991	117 (1.84)	82 (1.29)	89 (1.40)
1992	102 (1.60)	96 (1.51)	55 (0.86)
1993	134 (2.03)	116 (1.76)	64 (0.97)
1994	153 (2.02)	122 (1.61)	78 (1.03)
1995	124 (1.41)	113 (1.29)	98 (1.12)
1996	142 (1.51)	108 (1.15)	101 (1.07)

Cumulative growth rate (% p.a.)

(a) 1979–96	18.44	11.49	31.19
(b) 1979–88	32.40	17.40	44.22
(c) 1988–96	4.48	5.20	17.93

Note: Figures in brackets represent percentage share of total world export of this fish product.

Source: As in Table 6.1.

Thailand's master plan for development of fisheries for the period 1995–2000 has set a target of exporting not less than one million tons of fish products per annum and of increasing value added products by 10 per cent per annum. In order to maintain a world class quality of fish products, Thailand has moved towards setting up a product investigation laboratory in every coastal province for detection of antibiotic residue before harvest and sale to processing plants.

Thailand suffered a setback in marine capture fisheries between 1977 and 1980, primarily because of declaration of Exclusive Economic Zones (EEZs) by neighbouring countries like Malaysia, Vietnam, and Myanmar. This measure restricted the hitherto unlimited fishing operations by Thailand. In order to overcome this problem, Thailand adopted a two-pronged strategy. On the one hand, she placed a larger emphasis on aquaculture development (both mariculture and freshwater culture) and on the other hand, she began entering into Joint Venture agreements with other coastal states to operate in their EEZs. These twin strategies made Thailand recover the lost position to a large extent.

At the same time, Thailand invited considerable amounts of foreign investments, particularly from Japan, Hong Kong, the United States, and some European countries, to upgrade her technology not only in fish processing but also in farming, hatcheries, and feed mills. More than 50 new processing plants equipped with modern facilities were set up during 1990–3 for producing value added products. Thailand's skilled but relatively cheap labour added to her competitive advantage in the world market place.

Thailand has presence in almost all the major fish-consuming countries of the world. She presently exports to about 60 countries, though her major markets are Japan (35 per cent), the United States (33 per cent), and Europe (15 per cent). However, of late, Thailand has changed her strategy to focus more on the United States and other South East Asian countries than on Japan.

Thailand is also the largest exporter of canned tuna products. The raw materials for this product mainly comes from import—accounting for more than 70 per cent of tuna processing. The main types of tunas imported are skipjack followed by yellowfin tuna. Fifty per cent of the canning industry of Thailand is engaged in tuna processing and canning for exports. Some of these canneries have a capacity as large as 300 metric tons per day.

Speciality fish items produced by Thailand include fish sauce, fish paste, shrimp paste, surimi from threadfin bream, etc. The last fish item is gaining importance these days in Thailand. For this fish item the country has entered into a technical and marketing collaboration with Japan.

Presently, about 40 per cent of Thailand's export of fish products constitute processed, packed and other forms of fish preparations.

China

China entered the prepared and preserved fish products market only recently. She had a share of only 0.60 per cent of the world export in 1988, but within a span of only eight years she has emerged as the second largest exporter of this fish product in the world, capturing a market share of 12 per cent. We have earlier discussed at length China's policy initiative towards fishery development. As China's domestic fish consumption is high, she has deliberately chosen a policy to move towards high-value processed fish products, which she exports mainly to Japan and United States. Of late, China began exporting these products to the Middle East and some of the European countries too. This product, which constituted a mere 3.70 per cent share in the fish export basket of China in 1987 now claims a 40 per cent share of the basket. The export of value-added fish products is a new policy thrust of China.

Korea (R)

Korea (R) has increased her market share in this product to close to double that between 1979 and 1996. The share of this product in the total export of fish products from Korea (R) has also gone up from 6.78 per cent in 1979 to about 28 per cent in 1996. During the mid-period her growth rate in the export of this product has been phenomenal, next only to that of Thailand, but during the recent period there has been a gradual deceleration in the share of the world export as well as the cumulative growth rate. It appears that the growth in the export of this product is tapering off. This is primarily due to the rise in domestic consumption of fish on the face of a fall in catches due to overfishing, non-availability of skilled workers, and consequent high wages.

Korea (R) is known for producing high quality processed fish products but she is unable to bid for competitive prices because of the comparatively high cost of production. Due to the declining trend in both production and exports, a number of small companies have gone out of business while the larger companies are getting out of export of fishing and concentrating on domestic distribution. The trawler fishery, in which Korea (R) was the leader in the region, has lost the shine primarily due to the declaration of the EEZ and overfishing. Korea (R) had about 850 fishing vessels in the 1970s, which has came down to nearly 600 by now. A high interest rate and license fees are other reasons for the downfall. It is unlikely that Korea (R) will be able to maintain her present market share.

Japan

Japan was the leading exporter of prepared and preserved fish products of the world in the 1970s and the early part of the 1980s. But then, the market share began declining consistently. Between 1980 and 1996, Japan's share of the world export dropped by 88.21 per cent. The rate of growth has also been consistently negative throughout the period. The share of this product in the fish export basket of Japan has also declined from 62 per cent in 1979 to 38 per cent in 1996.

Japan is now a net importer of this fish product by a huge margin. In fact, she is the largest importer of this product in the world. Due to the increasing demand for this product for domestic consumption, Japan is hard-pressed to increase her share of the export. Presently, she exports certain speciality products of high value to the United States and some countries of Europe.

Morocco

Morocco emerged as a leading exporter of this fish product from Africa, with a share of about 3.5 per cent of the world export. But she is not able to maintain the market share in the face of stiff competition from the Asian exporters, though volume-wise her export has expanded by 128.57 per cent between 1979 and 1996. There is some recovery during the recent time (1988–96), which is indicated by a higher growth rate during the period as compared to the earlier period.

Philippines, Malaysia, Indonesia

These three countries are able to increase their share of export of this fish product, though volume-wise their exports are not at all comparable to that of the other Asian countries. All the three countries have recorded high cumulative growth rates in the export of this product for the period (1979–96). But there is a sharp fall in the growth rates of Philippines and Malaysia during the recent period (1988–96). Indonesia's growth rate has also fallen during the recent period but the fall is not as sharp as in the other two countries. The share of this product in the fish export basket of all these three countries has also gone up considerably. In the case of the Philippines, the share has increased from 8 per cent in 1979 to 33 per cent in 1996. In Malaysia, the increase is from 10.50 per cent in 1979 to about 34 per cent in 1996, while in Indonesia the growth is from a mere 0.45 per cent to 6 per cent during the same period. These three countries, being advanced developing nations of Asia with a high technological level in fishery production, are emerging as tough competitors in the world market of prepared and preserved fish products.

MARKET COMPETITIVE INDEX

Table 6.11 makes a comparative analysis of the market expansion characteristics of the countries discussed above.

Table 6.12 develops Market Competitive Indexes for all the countries discussed here who are engaged in the export of fish in prepared and preserved form.

TABLE 6.11

Comparative analysis of market expansion by select countries of Asia and Africa in prepared and preserved fish products

(per cent)

Country	Expansion of market	
	1979–96	1987–96
Thailand	2235.82	181.98
China	–	3041.67
Korea (R)	672.22	57.36
Japan	(–) 30.59	(–) 30.77
Morocco	128.57	87.23
Philippines	1675.00	136.67
Malaysia	535.29	50.00
Indonesia	10000.00	741.67
WORLD	332.57	116.48

EXPORT OF CRUSTACEANS, MOLLUSCS ETC. IN FRESH, FROZEN OR CHILLED FORM

CONTINENT ANALYSIS

Asia

Asia is the largest exporter of this group of fish products in the world, holding presently a market share of about 43 per cent, up from 36 per cent in 1979. But Asia is also a net importer of this product. We have seen earlier that in respect of import also, Asia holds the largest share (49.36 per cent as of 1996) of the market. This means that the flow of this product in and out of Asia is very high.

Prospects and Problems for Developing Nations

This is the second largest group of fish products coming immediately after fish in fresh, frozen or chilled form. Being a high value product the major consuming countries are in the developed world; in Asia, it is primarily Japan; in America, it is the United States; and in Europe, France and Spain. The major suppliers of this product, except Canada, belong to the developing

TABLE 6.12
Market Competitive Index and ranking of major countries engaged in the export of fish in prepared and preserved form

Country	Market Expansion Index (1987–96)	World market share (%) (1996)	Market volatility (1 – CV)	Comprehensive market parameter (col. 2 x 3 x 4)	Market Competitive Index	Country ranking
(1)	(2)	(3)	(4)	(5)	(6)	(7)
Denmark	0.43	0.0513	0.8852	0.0195	4.07	7
Netherlands	0.80	0.0354	0.7883	0.0223	4.65	6
Spain	1.63	0.0311	0.6434	0.0326	6.80	3
United States	0.99	0.0311	0.8274	0.0255	5.32	5
Canada	(–) 0.31	0.0282	0.9301	(–) 0.0081	(–) 1.69	21
Norway	(–) 0.06	0.0227	0.9250	(–) 0.0013	(–) 0.27	19
Iceland	5.18	0.0217	0.1472	0.0165	3.44	9
Chile	0.56	0.0193	0.8113	0.0088	1.84	14
France	1.07	0.0182	0.7076	0.0138	2.88	12
Ecuador	2.74	0.0160	0.3418	0.0150	3.13	11
United Kingdom	0.64	0.0149	0.8007	0.0076	1.59	15

(Contd.)

Table 6.12 (contd.)

Country (1)	Market Expansion Index (1987–96) (2)	World market share (%) (1996) (3)	Market volatility (1 – CV) (4)	Comprehensive market parameter (col. 2 x 3 x 4) (5)	Market Competitive Index (6)	Country ranking (7)
Greenland	0.32	0.0122	0.8710	0.0034	0.71	18
Portugal	0.38	0.0115	0.8522	0.0037	0.77	17
Thailand	1.56	0.1661	0.7728	0.2002	41.76	1
China	26.11	0.1201	0.0100	0.0314	6.55	4
Korea (R)	0.49	0.0443	0.8777	0.0191	3.98	8
Japan	(–) 0.26	0.0277	0.9094	(–) 0.0065	(–) 1.36	20
Morocco	0.75	0.0187	0.8502	0.0119	2.48	13
Philippines	1.17	0.0151	0.8527	0.0151	3.15	10
Malaysia	0.43	0.0115	0.8117	0.0040	0.83	16
Indonesia	6.37	0.0107	0.6582	0.0449	9.37	2
TOTAL	–	0.7278	–	0.4794	100.00	

world, many of whom have per capita income levels so low that there virtually does not exist any domestic market worth mentioning. The production of this product in these countries is invariably export oriented. While this may earn precious foreign exchange for these developing countries it also makes their seafood industry vulnerable to the vagaries of the export market. If this product is also the major component of export of total fish products, the vulnerability becomes more acute. An adverse movement in the international trade, be it economical, environmental, or political, may have the backlash effect of wiping out a major part of the industry, particularly in those countries where the industry is mainly composed of small scale units having low holding power due to inadequate net worth. One such country is India, to which we shall come later.

Lack of Bargaining Power

When a country produces a product solely for the export market, then besides suffering from the vagaries of the international market, she also loses the bargaining power in price realization, particularly in aquaproducts which are more perishable than any other food products. A large importing country would know that this exporting country does not have much option—the domestic market being virtually non-existent. Hence, they adopt arm-twisting techniques to bring down the price of this product by raising issues of quality etc., which, in many cases, may not really exist. The quality question is often found to be related to market demand. If the demand for the product is low the importing country raises the quality issues to dampen the price further. It is found that when demand becomes high, products of the same quality are picked up at a higher price. If the exporter has the financial power to wait and carry stock for the price to pick up, this arm-twisting would not pay off. But unfortunately, exporters from many developing nations mostly do not have this financial power.

The bargaining power of such developing countries is affected due to the lack of both physical and financial capacity to hold and carry stocks even within the biological limit of perishability. They are often compelled to unload stocks at unremunerative prices, akin to a distress sale.

Shrimp-orientation

Though Asia continues to be the net importer in this product (the margin being US$ 1045 million as of 1996), its expansion of the export market between 1979 and 1996 is 413.13 per cent, more than the world expansion of 330.84 per cent in the export of this product. Almost every fish producing developing nation in this region is tuned towards this export market. It would not be wrong to label the seafood industry in many of the countries as 'shrimp-oriented industry'. It is time to have a rethink about it considering the problems associated with such orientation, in the light of what we have discussed above.

During 1979–96, Asia's export grew by 10.10 per cent per annum. The cumulatively growth rate is higher (14.66 per cent per annum), if we consider 1979–88 period. But the growth has fallen to 5.19 per cent per annum during the recent period (1988–96) due to a fall in the world demand of this product, almost by the same percentage. However, the share of this product in the fish export basket of Asia has remained the same, at around 48 per cent during 1979–96.

America

America comes next in terms of the market share of this product. As of 1996, its share of the world export was 25.57 per cent, which is, however, down from 36.56 per cent in 1979. Unlike Asia, America is the net exporter of this product (the margin being US$ 654 million in 1996). The net export as a percentage of the total exports has gone up from 11.75 in 1979 to 14.75 in 1996.

As seen earlier, there is a general decline in the consumption of fish products in America particularly, among the developed Northern countries. It is also likely that a part of America is moving away from the consumption of this product as well, leaving an increasing level of output for the export market. America's compounded rate of growth in the export of this product is consistently over 6 per cent per annum, which has increased to more than 7 per cent during the recent period (1988–96).

Although America has not been able to fall in line with the rate of expansion of the world export market of this product, the continent is able to maintain a steady growth rate in export while, at the same time, containing the share of import of this product which has gone down from 27.70 per cent in 1979 to 22.03 per cent in 1996. The share of this product in the total fish exports from America has also gone up, though marginally from 39.94 per cent in 1979 to 40.60 per cent in 1996. During 1988–96, America's compounded rate of growth of import of this product has nearly halved (5.38 per cent per annum) from what it was during 1979–88 (10.28 per cent per annum). The trend is likely to continue and America will continue to make inroads into the export market of this product, though on a limited scale.

Europe

Europe's share of the export of this product has fallen from 15.44 per cent in 1979 to 12.75 per cent in 1996, though the share of this product in the fish export basket of the continent has remained at around 15 per cent during 1979–96. The continent is a net importer with a margin of US$ 2479 million, which is the highest among all the five continents. The net import is about 52 per cent of the total import. The largest importers are Spain and France. Unlike in America, the consumption of fish in general has gone up in Europe. For example, per capita consumption of EEC countries has gone up from 22.7 kg (average of 1988–90) to 23.1 kg during 1993–5, while for other Western European countries the figures are 14.9 and 27.3 kg, respectively. These countries are also in the most advanced part of Europe. Hence, they would ordinarily provide a good market for this high value fish product.

Europe's cumulative growth rate in the export of this fish product for the entire period is 7.75 per cent per annum; it is 9.67 per cent during 1979–88 and 5.63 per cent per annum during the recent period (1988–96). Although the growth rate has fallen during the recent period, the overall position appears to be steady. Quite a few countries are in the re-export trade, which contributes to the overall stability in the growth rate of exports. Otherwise, Europe will continue to remain as the net importer of this fish product.

Oceania

Oceania, a part of the developed world, presently (in 1996) commands a share of 5.17 per cent of the world export, down from 7.39 per cent in 1979. Though this small continent has a comparatively small share of the world export, it is a net exporter of this fish product; the net export being 82 per cent of the total export (with a margin of US$ 743 million as of 1996). The major exporters are Australia and New Zealand. Oceania's compounded growth rate in export is 6.70 per cent per annum for the entire period; the rate is found to have increased marginally to 7.01 per cent if we consider the 1979–88 period, but has also fallen marginally to 6.36 per cent if the recent period (1988–96) is considered. However, Oceania might be pursuing a strategy of reducing her export of crustaceans in fresh and frozen forms and increasing her export of fish in fresh, chilled or frozen form. This is borne by the fact that while the share of crustaceans etc. in the total fish exports from Oceania has come down from 69.04 per cent in 1979 to 47.34 per cent in 1996, the share of fish in fresh, frozen or chilled form has gone up from 23.16 per cent in 1979 to 43.48 per cent in 1996. This change might also have been caused due to an increase in the domestic consumption of crustaceans. The market expansion of Oceania between 1979 and 1996 is 201.32 per cent, which is below the expansion of the world market in this product. Oceania's per capita consumption of fish is exhibiting a decreasing tendency, but the available surplus might be increasing the export of fish in fresh, chilled or frozen form rather than the crustaceans in the same form. All these findings suggest that despite Oceania having presently a net export position in crustaceans products, it may not be able to hold on to its market share for long.

Africa

Africa is an emerging exporting country in fresh, frozen and chilled crustaceans and molluscs product. The continent has increased the market share consistently from 4.52 per cent in 1979 to 6.80 per cent in 1996. Ecuador is the leading exporter from Africa, and is also the sixth largest exporter of this product in the world. Africa's import is very small—about US$ 34 million in 1996. The net export position is 97.15 per cent of total export.

Africa's export grew at a formidable rate of 17.80 per cent per annum cumulatively during 1979–88. But during the recent period, the rate has come down substantially to 5.19 per cent per annum This is mainly due to a small export base of the continent with which it started in 1979, which gradually increased to a larger base in 1988. The market expansion of Africa between 1979 and 1996 is 548.36 per cent which is much above the world expansion of export in this fish product. The share of this product in the total fish exports from Africa has also increased from 38.25 per cent in 1979 to 41.12 per cent in 1996. All these findings indicate the emergence of Africa as a real competitor in the export market of this product.

TABLE 6.13

Comparative analysis of import–export performance of five continents in crustaceans, molluscs etc. in fresh, frozen or chilled form

(per cent)

Continents	World share of import		World share of export		Change in market share	
	1979	*1996*	*1979*	*1996*	*Import*	*Export*
Asia	52.50	49.36	36.09	42.98	(5.98)	19.10
America	27.70	22.03	36.56	25.57	(20.47)	(30.06)
Europe	18.91	26.93	15.44	12.75	42.41	(17.42)
Oceania	0.53	0.94	7.39	5.17	77.36	(30.04)
Africa	0.34	0.20	4.52	6.80	(41.18)	50.44

Note: Figures in brackets denote fall in market share.
Source: As in Table 6.1.

Table 6.13 summarizes the analyses of the import–export performance of the four continents.

Country Analysis

Table 6.14 analyses the export performance of the major exporting countries of America, Europe, and Oceania.

Countries of America, Europe, and Oceania

Canada

Canada is the largest exporter of shellfish in fresh, frozen or chilled form among all the countries of America, Europe, and Oceania. But the market share of the world export in this product has not improved much during the past 18 years. Canada is able to hold on to the market share by maintaining a consistent growth rate in export, in line with the expansion of the world market.

Canada is among the countries of America which is experiencing a decline in consumption of fish products. Her per capita consumption was 24.4 kg during 1987–9, which came down to 23.2 kg during 1993–5— a fall of about 5 per cent in six years which is quite substantial on a per capita basis. This must have contributed to Canada's maintaining the growth rate in export. The share of this product in the total fish exports of Canada has also gone up from 19 per cent in 1979 to 41.23 per cent in 1996.

Canada is a net exporting country in this fish product, the margin being US$ 491 million, which is about 52.5 per cent of the total export as on 1996. It is likely that Canada will be able to maintain her growth rate and market share, if not increase them in future.

United States

The United States was the leader in the export of shellfish in fresh, frozen or chilled form during 1979 and 1980. But then her share began falling continuously till 1986, when it started showing some signs of improvement. The rate of growth in export also improved, but she could never recover her lost position. Between 1979 and 1996, the United States lost 46.42 per cent of her market share. During the same period, the United States' export in absolute term has expanded by 120.54 per cent which is much below the expansion of world export in this product.

Unlike Canada, the United States' per capita consumption of fish products has remained virtually stagnant at around 21.5 kg since 1987, which indicates that the growth in consumption of fish has kept pace with the

TABLE 6.14
Exports of crustaceans, molluscs etc. in fresh, frozen or chilled form by major exporting countries of America, Europe and Oceania

(in million US dollars)

Year	Canada	United States	Ecuador	Australia	Mexico	Argentina	Spain
1979	210 (5.16)	336 (8.25)	31 (0.76)	248 (6.09)	372 (9.13)	55 (1.35)	201 (4.93)
1980	243 (6.33)	271 (7.06)	72 (1.88)	216 (5.63)	398 (10.36)	14 (0.36)	129 (3.36)
1981	255 (6.22)	267 (6.52)	93 (2.27)	285 (6.96)	357 (8.72)	18 (0.44)	148 (3.61)
1982	278 (5.95)	253 (5.41)	123 (2.63)	361 (7.73)	453 (9.69)	45 (0.96)	98 (2.10)
1983	313 (6.22)	196 (3.90)	239 (4.75)	310 (6.16)	381 (7.57)	81 (1.61)	121 (2.41)
1984	282 (6.13)	160 (3.48)	161 (3.50)	285 (6.20)	430 (9.35)	108 (2.35)	145 (3.15)
1985	345 (7.06)	163 (3.33)	157 (3.21)	291 (5.95)	338 (6.91)	57 (1.17)	157 (3.21)
1986	423 (6.62)	253 (3.96)	307 (4.81)	289 (4.52)	392 (6.14)	63 (0.99)	153 (2.39)
1987	485 (6.04)	376 (4.68)	385 (4.80)	371 (4.62)	470 (5.85)	60 (0.75)	152 (1.89)
1988	495 (4.49)	471 (4.27)	388 (3.52)	431 (3.91)	400 (3.63)	88 (0.80)	166 (1.50)
1989	427 (4.04)	529 (5.01)	329 (3.11)	365 (3.46)	386 (3.65)	94 (0.89)	168 (1.59)
1990	497 (4.22)	725 (6.16)	341 (2.90)	389 (3.30)	241 (2.05)	69 (0.59)	180 (1.53)
1991	469 (3.60)	837 (6.42)	493 (3.78)	425 (3.26)	258 (1.98)	100 (0.77)	195 (1.50)

(Contd.)

Table 6.14 (contd.)

Year	Canada	United States	Ecuador	Australia	Mexico	Argentina	Spain
1992	588 (4.66)	852 (6.75)	529 (4.19)	497 (3.94)	240 (1.90)	219 (1.73)	214 (1.70)
1993	677 (5.09)	845 (6.36)	468 (3.52)	539 (4.05)	326 (2.45)	331 (2.49)	238 (1.79)
1994	888 (5.44)	875 (5.36)	557 (3.41)	624 (3.82)	376 (2.30)	292 (1.79)	324 (1.98)
1995	1064 (6.04)	790 (4.48)	679 (3.85)	659 (3.74)	537 (3.05)	320 (1.82)	318 (1.81)
1996	936 (5.33)	741 (4.42)	631 (3.60)	622 (3.54)	574 (3.27)	433 (2.47)	431 (2.46)
Cumulative growth rate (% p.a.)							
(a) 1979–96	9.19	4.76	19.39	5.56	2.58	12.91	4.59
(b) 1979–88	10.00	3.82	32.42	6.33	0.81	5.36	(–) 2.10
(c) 1988–96	8.29	5.83	6.27	4.69	4.62	22.04	12.67

Year	United Kingdom	Denmark	New Zealand	The Netherlands	France	Italy	Brazil
1979	91 (2.23)	60 (1.47)	43 (1.06)	82 (2.01)	54 (1.33)	22 (0.54)	114 (2.80)
1980	102 (2.66)	86 (2.24)	71 (1.85)	90 (2.34)	68 (1.77)	22 (0.57)	90 (2.34)
1981	99 (2.42)	89 (2.17)	59 (1.44)	74 (1.81)	57 (1.39)	22 (0.54)	104 (2.54)
1982	93 (1.99)	82 (1.75)	84 (1.80)	62 (1.33)	66 (1.41)	20 (0.43)	122 (2.61)
1983	102 (2.03)	102 (2.03)	87 (1.73)	71 (1.41)	77 (1.53)	21 (0.42)	99 (1.97)
1984	105 (2.28)	103 (2.24)	105 (2.28)	60 (1.31)	67 (1.46)	27 (0.59)	148 (3.22)

(Contd.)

Table 6.14 (contd.)

Year	United Kingdom	Denmark	New Zealand	The Netherlands	France	Italy	Brazil
1985	110 (2.25)	123 (2.52)	100 (2.05)	61 (1.25)	80 (1.64)	28 (0.57)	141 (2.88)
1986	164 (2.57)	193 (3.02)	102 (1.60)	93 (4.56)	126 (1.97)	40 (0.63)	119 (1.86)
1987	234 (2.91)	284 (3.54)	140 (1.74)	125 (1.56)	160 (1.99)	42 (0.52)	143 (1.78)
1988	230 (2.09)	292 (2.65)	104 (0.94)	134 (1.21)	180 (1.63)	52 (0.47)	147 (1.33)
1989	251 (2.38)	289 (2.74)	152 (1.44)	137 (1.30)	186 (1.76)	60 (0.57)	95 (0.90)
1990	316 (2.68)	310 (2.63)	125 (1.06)	173 (1.47)	216 (1.84)	76 (0.65)	111 (0.94)
1991	337 (2.59)	316 (2.42)	132 (1.01)	164 (1.26)	191 (1.47)	86 (0.66)	119 (0.91)
1992	293 (2.32)	280 (2.22)	152 (1.20)	194 (1.54)	193 (1.53)	87 (0.69)	117 (0.93)
1993	252 (1.90)	285 (2.14)	149 (1.12)	207 (1.56)	182 (1.37)	87 (0.65)	123 (0.93)
1994	305 (1.87)	314 (1.92)	214 (1.31)	232 (1.42)	184 (1.13)	98 (0.60)	124 (0.76)
1995	360 (2.04)	331 (1.88)	234 (1.33)	237 (1.35)	214 (1.21)	105 (0.60)	111 (0.63)
1996	371 (2.11)	386 (2.20)	254 (1.45)	232 (1.32)	203 (1.16)	117 (0.67)	89 (0.51)

Cumulative growth rate (% p.a.)

	United Kingdom	Denmark	New Zealand	The Netherlands	France	Italy	Brazil
(a) 1979–96	8.62	11.57	11.01	6.31	8.10	10.33	(–) 1.45
(b) 1979–88	10.65	19.22	10.31	5.61	14.31	10.03	2.87
(c) 1988–96	6.16	3.55	11.81	7.10	1.51	10.67	(–) 6.08

Note: Figures in brackets represent percentage share of total world export of this fish product.
Source: **As in** Table 6.1.

growth in population, the latter having grown by 7.50 per cent between 1987 and 1996. This means that the absolute quantitative level of demand has increased at least by the above percentage, which must be met from imports when domestic production is falling. Being a high-income developed country, the demand in the United States for this high value crustacean product would also be high. Against this, the United States suffered from a negative rate of growth in catches during 1988–96. All these have made the United States a net importer of this fish product, the margin being US$ 2568 million as on 1996. Her net import is about 78 per cent of the total import. The share of this product in the total export of fish products from the United States has also fallen from 31.17 per cent in 1979 to 24.24 per cent in 1996. Under the given circumstances it is unlikely that the United States will be able to increase her market share any further, rather it may fall in the years to come.

Mexico, Brazil, Argentina

Among the four South American countries featured in Table 6.14, Mexico and Brazil have lost their share of the world export considerably, while Argentina has improved its share. It may be noted that Brazil's total export of fish products have also come down from US$ 143 million in 1979 to US$ 123 million in 1996. All these three countries have very little import of this product. Their major export markets are in the United States, followed by Japan. Hence, the fortunes of their exports depend to a large extent on the import demands from these two countries. Both Mexico and Brazil are suffering from severe deceleration in the growth rate of their catches both in marine and inland waters, which might have contributed towards the sharp fall in their share of exports of this fish product. Although the product continues to have the largest share of export of all fish products from these countries, its share has fallen in Brazil from 79.72 per cent in 1979 to 72.35 per cent in 1996 and in Mexico from 87.94 per cent to 74.06 per cent between the respective years.

Argentina has considerably improved her nominal catches over the years, with a healthy growth rate both in marine and inland waters, as seen earlier. This must have enabled the country to increase her market share in this fish product. Argentina's share of the world export of this product has increased by about 83 per cent between 1979 and 1996. The total expansion in her export is 687.27 per cent during the same period, which is more than double the expansion rate of world export. This is due to the high cumulative growth rate of 22 per cent per annum in export during the recent period (1988–96). The share of this product in Argentina's total fish exports has also gone up from 26.96 per cent in 1979 to 43.08 per cent in 1996. Considering all these indications we may say that Argentina will soon emerge as the leading exporter of this fish product in the entire continent.

Ecuador

This country deserves a special mention. From a humble beginning she has now emerged as the sixth largest exporter of crustacean products in the world, having increased her share of world export by 374 per cent between 1979 and 1996. Massive investment has been made in aquaculture development, aided by North American countries and multilateral organizations. Although during the recent period, the growth rate in her export has decelerated, it is likely that she will recover from it soon. Domestic consumption of fish is on the lower side in Ecuador. There is also a decline in her per capita consumption of fish—from 8.8 kg during 1987–9 to 7.2 kg during 1993–95. This has helped boost up her export of this high value aquatic product. She is now going to stay as a major contender in the export market of fresh, frozen or chilled crustaceans and molluscs products.

Australia, New Zealand

Of the two countries of Oceania, Australia has lost her share of export of crustaceans etc. between 1979 and 1996 by about 42 per cent, while New Zealand's share has improved by 37 per cent approximately. Both the two countries are, however, net exporters in this product. In the case of Australia, the net margin is US$ 479 million, constituting 77 per cent of the total export as on 1996. In the case of New Zealand, the margin is US$ 242 million, which is more than 95 per cent of the total export of the same year. New Zealand's nominal catch of fish is on the rise while in Australia it is decreasing. New Zealand's per capita consumption of all fish has fallen by about 20 per cent between 1987 and 1985; in Australia the per capita consumption has increased by about 2 per cent during the same period. All these have contributed to the decline in Australia's market share and increase in New Zealand's share.

Spain

In Europe, Spain is the market leader in the export of crustaceans etc., followed closely by the United Kingdom and Denmark. However, while the United Kingdom was able to maintain her share of the export in this product and Denmark could improve her share, Spain's share has decreased by more than 50 per cent between 1979 and 1996. The share of this product in Spain's total export of fish products has also come down from 49.26 per cent in 1979 to 30 per cent in 1996.

Spain is also a net importer by a margin of US$ 856 million, constituting about 66.50 per cent of the total import of this product as on 1996. Spain's consumption of fish products is among the highest in Europe. Her per capita consumption has also gone up from 37.1 kg during 1987–9 to 42.7 kg during 1993–5, that is, a rise of more than 15 per cent between the two periods, while the rise in population is only 1.52 per cent during the same period. This indicates that an additional demand for fish product has been created in the country. Spain is also suffering consistently from a negative growth rate in fish production. Considering all these factors, it can be concluded that Spain's market share is going to further decline in the future.

Denmark and United Kingdom

In contrast to Spain, both the United Kingdom and Denmark are net exporters of this fish product. Their net exports of this product are 39 per cent and 25 per cent of their respective total exports. Denmark's share of world export of crustaceans etc. showed a marked increase during the early 1980s but it showed a declining trend thereafter. However, as compared to 1979, by 1996 she experienced an expansion of about 50 per cent in the share of her export. The share of this product in the total export of fish products from Denmark has also increased from 8.5 per cent in 1979 to 18.18 per cent in 1996. This may be at the cost of a decrease in the share of fish in fresh, frozen and chilled form in the fish export basket of Denmark, which reduced from 66.34 per cent in 1979 to 45.45 per cent in 1996. It is likely that despite a fall in the growth rate of export of this product during the recent period, Denmark is likely to continue as a major exporter of crustaceans, molluscs etc. in fresh, frozen or chilled form.

The United Kingdom has, however, maintained her average share of the world export of crustaceans, molluscs etc. in fresh, frozen or chilled form. The share of this product in the fish export basket of the United Kingdom has also gone up marginally from 29.26 per cent in 1979 to 33.40 per cent in 1996.

There is also an element of re-export in the total exports of this fish product by these two countries.

The Netherlands

This is another country whose export and import of this fish product is almost the same, which suggests a strong element of re-export content in their international trade. The share of this product in the country's total exports has also remained the same throughout the period, at about 16.6 per cent during 1979–96.

France and Italy

The share of the world export of this product of both France and Italy is small but more or less steady over past 18 years. Both the two countries are net importers with a sizeable margin. France's net import margin is US$ 688 million, which constitutes more than 77 per cent of the total import, while the corresponding figures for Italy are US$ 685 million and nearly 85.5 per cent, respectively.

Asian Countries

Table 6.15 analyses the performance of the major exporting countries of Asia.

TABLE 6.15

Export of crustaceans, molluscs etc. in fresh, chilled or frozen form by major exporting countries of Asia

Year	Thailand	Indonesia	India	Vietnam	Korea (R)
1979	223 (5.48)	205 (5.03)	301 (7.39)	–	174 (4.27)
1980	192 (5.00)	184 (4.79)	226 (5.88)	–	166 (4.32)
1981	201 (4.91)	174 (4.25)	309 (7.54)	–	197 (4.81)
1982	249 (5.33)	193 (4.13)	365 (7.81)	–	218 (4.67)
1983	263 (5.23)	204 (4.06)	347 (6.90)	–	235 (4.67)
1984	255 (5.55)	203 (4.42)	289 (6.29)	–	230 (5.00)
1985	273 (5.58)	208 (4.25)	307 (6.28)	–	224 (4.58)
1986	397 (6.21)	297 (4.65)	352 (5.51)	–	328 (5.13)
1987	508 (6.33)	369 (4.60)	463 (5.57)	–	456 (5.68)
1988	642 (5.82)	527 (4.78)	396 (3.59)	407 (3.69)	537 (4.87)
1989	923 (8.74)	574 (5.43)	457 (4.33)	483 (4.57)	424 (4.01)
1990	1061 (9.01)	710 (6.03)	468 (3.98)	436 (3.70)	366 (3.11)
1991	1355 (10.40)	796 (6.11)	513 (3.94)	305 (2.34)	411 (3.15)
1992	1566 (12.40)	789 (6.25)	570 (4.52)	257 (2.04)	373 (2.95)
1993	1826 (13.73)	907 (6.82)	700 (5.27)	347 (2.61)	328 (2.47)
1994	2326 (14.24)	1051 (6.43)	939 (5.75)	422 (2.58)	327 (2.00)
1995	2412 (13.69)	1081 (6.14)	798 (4.53)	424 (2.41)	399 (2.27)
1996	2373 (13.52)	1064 (6.06)	1025 (5.84)	380 (2.17)	373 (2.13)

(Contd.)

Table 6.15 (contd.)

Cumulative growth rate (% p.a.)

(a) 1979–96	14.92	10.17	7.47	–	4.59
(b) 1979–88	12.47	11.06	3.09	–	13.34
(c) 1988–96	17.75	9.18	12.62	(–) 0.85	(–) 4.45

Year	Bangladesh	Philippines	Singapore	Malaysia	Japan
1979	41 (1.01)	45 (1.10)	36 (0.88)	136 (1.34)	51 (1.25)
1980	36 (0.94)	28 (0.73)	26 (0.68)	67 (1.74)	69 (1.80)
1981	36 (0.88)	32 (0.78)	31 (0.76)	51 (1.25)	62 (1.51)
1982	61 (1.31)	42 (0.90)	42 (0.90)	53 (1.13)	62 (1.33)
1983	66 (1.31)	57 (1.13)	48 (0.95)	80 (1.59)	108 (2.15)
1984	76 (1.65)	43 (0.94)	54 (1.17)	50 (1.09)	87 (1.89)
1985	90 (1.84)	75 (1.53)	51 (1.04)	45 (0.92)	88 (1.80)
1986	123 (1.93)	119 (1.86)	65 (1.02)	60 (0.94)	122 (1.91)
1987	122 (1.52)	177 (2.20)	97 (1.21)	79 (0.98)	119 (1.48)
1988	120 (1.09)	271 (2.46)	113 (1.02)	97 (0.88)	87 (0.79)
1989	143 (1.35)	273 (2.58)	127 (1.20)	101 (0.96)	98 (0.93)
1990	134 (1.14)	257 (2.18)	150 (1.27)	115 (0.98)	74 (0.63)
1991	155 (1.19)	319 (2.45)	208 (1.60)	140 (1.07)	79 (0.61)
1992	142 (1.12)	258 (2.04)	193 (1.53)	142 (1.12)	70 (0.55)
1993	184 (1.38)	282 (2.12)	179 (1.35)	123 (0.93)	78 (0.59)
1994	241 (1.48)	310 (1.90)	215 (1.32)	144 (0.88)	93 (0.57)
1995	234 (1.33)	291 (1.65)	193 (1.10)	159 (0.90)	96 (0.54)
1996	254 (1.45)	220 (1.25)	169 (0.96)	151 (0.86)	124 (0.71)

Cumulative growth rate (% p.a.)

(a) 1979–96	11.32	9.78	9.52	0.62	5.37
(b) 1979–88	12.67	22.08	13.55	(–) 3.69	6.11
(c) 1988–96	9.83	(–) 2.57	5.16	5.69	4.53

Note: Figures in brackets represent percentage share of total world export of this fish product.
Source: As in Table 6.1.

Thailand

Thailand is the leader in the world export market of crustaceans, molluscs etc. in fresh, frozen or chilled form. This product also occupies the first place in the total export of fish products from Thailand. She has increased her share in the world export market by 147 per cent between 1979 and 1996 by maintaining a steady growth rate of around 15 per cent per annum.

We have discussed earlier the prospects and problems faced by Thailand in the matter of fish production and exports. It should be added here that the major items of crustaceans export of Thailand compose of shrimps and prawns. With the declaration of the EEZ and continuous overfishing, Thailand's marine landings of these products came down (though some amount of poaching still continues). In order to compensate for this fall, Thailand went in for intensive aquaculture in shrimp production. The contribution of aquaculture to total shrimp–prawn production has increased from 15 per cent in 1986 to about 75 per cent in 1996. The production of Black Tiger and King Prawn, having a high commercial value, has gone up substantially during the recent years. The majority of shrimp farms in Thailand are small in size, but are highly efficient. However, during the 1990s aquaculture in Thailand ran into severe environmental problems. As a result, Thailand has restricted the expansion of hectarage while placing larger emphasis on coastal aquaculture. The overall picture that emerges is that Thailand's production of fish products is on the decline. Particularly hit is the export of crustaceans in fresh, frozen or chilled form, whose share of world export has stagnated to around 13 per cent during the past five years. The share of this product in the total export of fish products from Thailand has decreased substantially, from 71.25 per cent in 1979 to 54.24 per cent in 1996. Owing to the declining trend in fish production, the country has moved towards a strategy of developing the export of fish with high value addition in prepared and preserved form. As a result, the share of this product in the total fish exports of Thailand has gone up from 21.40 per cent in 1979 to 35.17 per cent in 1996.

Thailand has, however, a way of making a recovery. It is likely that she might lose her market dominance in the export of crustaceans in fresh, frozen or chilled form, but this can be compensated by increasing the share of high value speciality fish products.

Indonesia

Indonesia occupies the second largest position in the world export of this fish product. Her share of the world export came to be stabilized from 1989, and since then she has been holding on to the market share of around 6.5 per cent by maintaining a steady rate of growth in exports.

Earlier we have discussed at length the emphasis placed by the government on aquaculture development, which has resulted in the growth of the crustaceans export of this country. Between 1979 and 1996, Indonesia's export has expanded by 419 per cent, much above the world expansion of 331 per cent during the same period. But between 1987 and 1996 the rate of expansion has come down to 188 per cent, close to the world expansion level of 119 per cent. As in Thailand, the share of this product in the total fish exports of Indonesia has come down from 92.76 per cent in 1979 to 63.41 per cent in 1996. Percentage-wise, the fall is about 8 points higher than in Thailand. Indonesia has also put emphasis on prepared and preserved fish products, whose share in the total fish exports has gone up from a mere 0.45 per cent in 1979 to 6.02 per cent in 1996. But simultaneously, she has also stepped up her export of fish in fresh, frozen or chilled form whose share has increased from 4.53 per cent in 1979 to 25.33 per cent in 1996.

Indonesia does not have the resilience of Thailand to make a recovery on her own. She has adopted a rather softer strategy of increasing the export of fish in fresh, frozen or chilled form more than that of the prepared and preserved form (the latter also calls for high capital investment) in the face of stagnation in the export of crustaceans in fresh, chilled or frozen form.

Vietnam

Recent data indicate that Vietnam is presently suffering from a deceleration in the growth of exports of this fish product. Her share of the world export of this fish product export has also fallen by more than 40 per cent between 1988 and 1996. Vietnam is likely to have reached a plateau in her fish catches, which may be responsible for the fall in the market share. Of late, she has entered into joint ventures with some of the advanced developing nations of Asia to foster a growth in catches.

India

India, which was the third largest supplier of crustaceans and molluscs etc. product to the world in 1979, continues to hold the same position even today, though she has lost more than 21 per cent of her share to Thailand and Indonesia. This product accounts for around 87 per cent of the total fish exports of India since 1979. India's case is discussed in a separate section.

Korea (R), Malaysia, and Singapore

Korea(R) and Malaysia are net exporters, having a margin of US$ 119 million and US$ 80 million, respectively, which constitute about 32 per cent and 53 per cent of their respective total exports. It appears that both the two countries have high re-export contents in their exports. Similar is the case with Singapore, though she is a net importer with a margin of US$ 25 million that account for about 25 per cent of the total import of this fish product by the country.

There is a marginal increase in the share of this product in the total fish exports of Korea(R). In 1979 the share was 21.86 per cent, which increased to 25 per cent in 1996. However, in the case of Malaysia the share has fallen drastically, from 83.43 per cent in 1979 to 47.34 per cent in 1996. During the same time, the share of prepared and preserved fish products has gone up from 10.43 per cent to 33.86 per cent, signifying a change in emphasis on the latter kind. In the case of Singapore, the share of this product in the total fish exports has fallen marginally, from 36.73 per cent in 1979 to 31.30 per cent in 1996.

Countries who are re-exporters of this fish product are experiencing a decline in their growth rates. The primary reason behind this is that many of the Asian countries who were hitherto making exports to these countries have started exporting directly.

Bangladesh

Bangladesh is emerging as an important exporter of crustaceans etc. in fresh or frozen form. Her rate of growth has also been impressive. The coastal areas of the Bay of Bengal and a large number of rivers and ponds in Bangladesh have been made use of extensively for both freshwater and brackish water shrimp production. Bangladesh presently holds the fifth position in inland water fish production of the world, with a share of about 5 per cent of global inland water catches. This country, like many other Group B developing nations, does not have much of a domestic market for this high value fish item. These are produced mainly for the export market.

Philippines

This is another country which has a well developed aquaculture. She comes immediately after Bangladesh in respect of the share of the world inland water fish production. Between 1975 and 1992, the inland water catches of the Philippines expanded by 171.50 per cent—more than the world expansion of 123.80 per cent during the period. Philippines' export of shellfish grew steadily from 1979 to 1993. But since then it is showing signs of deceleration, as indicated by a negative growth rate. The domestic consumption of fish in the Philippines is very high. Per capita consumption has remained close to 34 kg since 1987, against a rise in population by 11.66 per cent between 1987 and 1995. This means that there has been an expansion of the demand of fish domestically in absolute terms. This has happened at a time when the total nominal catches of the Philippines has fallen drastically, the growth rate being a mere 0.92 per cent. In particular, the growth in inland water catches, which is the main source of crustaceans export, has gone down to a mere 1 per cent per annum. Although it is unlikely that the high value crustaceans product has much of a domestic market in Philippines, the general expansion of domestic demand and contraction in nominal catches, particularly inland water catches, have contributed to the fall in the market share and growth in exports of this product.

Japan's export is the smallest as compared to the other countries discussed here for crustaceans etc. She is a net importer of this fish product by a huge margin of US$ 6562 million, which accounts for about 98 per cent of the import of crustaceans etc.

MARKET COMPETITIVE INDEX

Table 6.16 analyses the comparative market expansion in this product by the countries discussed above.

TABLE 6.16
Comparative expansion of export of crustaceans, molluscs etc.in fresh, frozen or chilled form by major exporting countries of the world

(per cent)

Country	Market expansion	
	1979–96	*1987–96*
Canada	345.71	92.99
United States	120.54	97.07
Ecuador	1935.48	63.90
Australia	150.80	67.65
Mexico	54.30	22.13
Argentina	687.27	621.67
Spain	114.43	183.55
United Kingdom	307.69	58.55
Denmark	543.33	35.92
New Zealand	490.70	81.43
The Netherlands	182.93	85.76
France	275.93	26.88

(Contd.)

Table 6.16 (contd.)

	Market expansion	
Country	1979–96	1987–96
Italy	431.82	178.57
Brazil	(–) 21.93	(–) 37.76
Thailand	964.13	367.13
Indonesia	419.02	188.35
India	240.53	121.38
Vietnam	–	(–) 6.64
Korea(R)	114.37	(–) 18.20
Bangladesh	519.51	108.20
Philippines	388.89	24.29
Singapore	369.44	74.23
Malaysia	11.03	91.13
Japan	143.14	4.20
WORLD	330.84	118.56

Table 6.17 shows the Market Competitive Indexes for the major exporting countries of the world in crustaceans, molluscs etc. in fresh, chilled or frozen form.

EXPORT OF FISH IN SALTED, DRIED OR SMOKED FORM

We have seen earlier that this 'ancient' form of fish product is predominantly traded among the developed nations of the world. Although sun drying with or without salt continues to be an important method of product preparation, certain technological advancement has taken place in the methods of production as well as presentations. Instead of the whole fish, some products are salted, dried or smoked in fillet form. The major products which come within this group are fish livers and roes, dried, salted, smoked or in brine; herrings and salmons whole or in fillets in smoked form; cods, dried and or salted; herrings, anchovies, and cods, salted or in brine; oysters and scallops, dried, salted or in brine; mussels, dried or salted; octopus, dried or salted, etc. Markets in this product group differ significantly in terms of fishes, their preparation and presentation. This fish product is highly culture specific.

Among the four groups of fish products, this group has the smallest share

TABLE 6.17

Market Competitive Indexes for major exporting countries of the world in the international trade of crustaceans, molluscs etc. in fresh, frozen or chilled form

Country	Market Expansion Index (1987–96)	Market share (%) (1996)	Degree of fluctuations (1 – CV)	Comprehensive market parameter (col. 2 x 3 x 4)	Market Competitive Index	Country ranking
(1)	(2)	(3)	(4)	(5)	(6)	(7)
Canada	0.78	0.0533	0.6705	0.0279	4.67	6
United States	0.82	0.0442	0.8140	0.0295	4.93	5
Ecuador	0.54	0.0360	0.7640	0.0148	2.48	8
Australia	0.57	0.0354	0.7939	0.0160	2.67	7
Mexico	0.19	0.0327	0.6903	0.0043	0.72	17
Argentina	5.24	0.0247	0.4165	0.0539	9.02	3
Spain	1.55	0.0246	0.6551	0.0250	4.19	6
United Kingdom	0.49	0.0211	0.844	0.0087	1.46	10
Denmark	0.30	0.0220	0.9011	0.0059	0.99	14
New Zealand	0.69	0.0145	0.7066	0.0071	1.19	13
Netherlands	0.72	0.0132	0.8003	0.0076	1.27	12

(Contd.)

Table 6.17 (contd.)

Country	Market Expansion Index (1987–96)	Market share (%) (1996)	Degree of fluctuations (1 – CV)	Comprehensive market parameter (Col.2 x 3 x 4)	Market Competitive Index	Country ranking
(1)	(2)	(3)	(4)	(5)	(6)	(7)
France	0.23	0.0116	0.9341	0.0025	0.42	18
Italy	1.51	0.0067	0.7724	0.0078	1.31	11
Brazil	(–)0.32	0.0051	0.8612	(–) 0.0014	(–) 0.23	22
Thailand	3.10	0.1352	0.6091	0.2553	42.74	1
Indonesia	1.59	0.0606	0.7643	0.0736	12.32	2
India	1.02	0.0584	0.6731	0.0401	6.71	4
Vietnam	(–)0.06	0.0217	0.8270	(–) 0.0011	(–) 0.18	21
Korea(R)	(–)0.15	0.0213	0.8479	(–) 0.0027	(–) 0.45	23
Bangladesh	0.91	0.0145	0.7278	0.0096	1.61	9
Philippines	0.20	0.0125	0.8980	0.0022	0.37	19
Singapore	0.63	0.0096	0.8054	0.0049	0.82	16
Malaysia	0.77	0.0086	0.8396	0.0056	0.94	15
Japan	0.04	0.0071	0.8246	0.0002	0.03	20
TOTAL	–	–	–	0.5973	100.00	

in international trade—about 5.70 per cent as on 1996, down from 8.18 per cent in 1979—but its cumulative growth rate at 6.54 per cent, though lower than the growth rate of all the groups combined together, cannot be said to be negligible. Its total export market has also expanded by 193.41 per cent between 1979 and 1996. Despite a trend of deceleration in the market share (which will continue), it is very unlikely that this product will go out of the fish market in absolute terms.

CONTINENT ANALYSIS

Europe

Europe is the largest exporter of this fish product, holding a 57.74 per cent share of the world market. We may recall that Europe is also the largest importer of this product, holding a share of about 49 per cent of the world import. However, the overall position of Europe is that the continent is a net exporter with a margin of US$ 220 million which account for 13.37 per cent of the total exports. On the face of it, it can be concluded that there exists a large element of re-export, but as we shall see later, the European countries which are engaged mainly in the import of this product are not the same as those countries who make the exports.

Europe's share of the world export of this product has decreased by 11.3 per cent between 1979 and 1996. The market expansion during this period is 160.28 per cent, as against a world expansion of 193.41 per cent. Europe's export has grown at a cumulative rate of 5.79 per cent per annum during 1979–96. However, during the recent period of 1988–96, the growth rate has fallen to 4.90 per cent per annum, as compared to 6.60 per cent per annum during 1979–88. The share of this product in the total fish products export from Europe has also gone down from 15.05 per cent in 1979 to 11.26 per cent in 1996. It appears that Europe is gradually increasing its export of fish in fresh, chilled or frozen form and decreasing the export of fish in salted, dried or smoked form.

Part of Europe's market share has been taken over by Asia, but despite this, all the other indications suggest that Europe will continue to be the market leader of this product for some years to come.

America

America is the second largest exporter of salted, dried and smoked fish products. This continent, like Europe, is also a net exporter, with a margin of US$ 89 million that constitutes 16.3 per cent of total export of this product by America. Here also, except for the United States whose import and export of this product closely match, the import or export characteristics are distinct for the other countries. In the case of America also, the share of the world export has fallen from 24 per cent in 1979 to 19.16 per cent in 1996—a fall of about 20 per cent. The share of this product in the fish export basket of America has also fallen from 6.25 per cent in 1979 to 4.94 per cent in 1996. America's cumulative growth rate in export is higher than that of Europe at 9.97 per cent per annum. But between 1988 and 1996, America has registered a negative growth rate of 0.05 per cent per annum.

It appears that America's growth in export of this product is stagnating, and this trend is likely to continue in the future. Hence, it is expected that America's share of the world export in this product will fall further.

Asia

Asia has made significant progress in the export of dried, smoked or salted fish products. The market share of Asia has more than doubled between 1979 and 1996—from 8.03 per cent in 1979 to 16.85 per cent in 1996. Asia's cumulative rate of growth for the entire period of 1979–96 is 11.28 per cent per annum, but when we consider the recent period (1988–96) we find that it is 9.22 per cent per annum. The market expansion of Asia is the highest among all the continents. It is 580.23 per cent between 1979 and 1996—more than two and a half times the world expansion of export in this product. The share of salted, dried or smoked form of fish in the total fish exports from Asia has also increased from 2.57 per cent in 1979 to 3.03 per cent in 1996. However, we must recall that Asia is a net importing continent in this product, with a margin of US$ 441 million which accounts for 47.88 per cent of the import. The two major exporting countries from Asia are, to a large extent, re-exporters of this fish product, as will be seen later.

It was seen earlier that Japan is the world's largest importer of this product. However, her export is quite negligible.

Considering the trends mentioned above there is all likelihood that Asia will overtake Europe in her share of world export of this fish product.

Africa and Oceania

Presently, Africa and Oceania have a very small share of the world exports of this product. Their imports are also very negligible.

Africa has substantially reduced her share of the world export, from 21.39 per cent in 1979 to 5.08 per cent in 1990. The share of this product in the total fish export basket of Africa has also fallen from 5.41 per cent in 1979 to 1.52 per cent in 1996. Africa has made a major thrust towards increasing the export of fish in fresh, frozen or chilled form, whose share in the total fish products export of the continent has gone up from 22.86 per cent in 1979 to

33.16 per cent in 1996. But the share of shellfish in fresh, frozen or chilled form has remained at around 40 per cent during the whole period. The export of fish in salted, dried or smoked form is of low priority now.

Oceania is not an exporter of much significance. The rise in the export of this product is only about US$ 7 million during the past 18 years.

Table 6.18 summarizes the export–import performance of the five continents in this fish product.

TABLE 6.18

Comparative analysis of export–import performance of five continents in dried, salted or smoked fish products

(per cent)

Continents	Share of world import		Share of world export		Change in market share	
	1979	1996	1979	1996	Import	Export
Asia	40.16	31.62	8.03	16.85	(21.26)	109.84
America	13.88	15.69	24.00	19.16	13.04	(20.17)
Europe	34.27	48.92	65.09	57.74	42.75	(11.29)
Oceania	0.86	0.62	0.21	0.74	(27.90)	252.38
Africa	8.71	1.41	2.68	1.54	(83.81)	(42.54)

Note: Figures in brackets denote fall in market share.
Source: As in Table 6.1.

Table 6.19 presents the export performance data of the major exporting countries of America and Europe.

TABLE 6.19

Export of fish in salted, dried or smoked form by major exporting countries of America and Europe

(in million US dollars)

Year	Norway	Denmark	Canada	Iceland
1979	270 (27.81)	52 (5.36)	66 (6.80)	131 (13.49)
1980	322 (27.04)	63 (5.29)	93 (7.81)	225 (18.89)
1981	368 (24.97)	64 (4.34)	127 (8.62)	298 (20.22)
1982	284 (23.91)	70 (5.89)	124 (10.44)	181 (15.24)

(Contd.)

Table 6.19 (contd.)

Year	Norway	Denmark	Canada	Iceland
1983	236 (22.78)	66 (6.37)	95 (9.17)	149 (14.38)
1984	209 (22.33)	64 (6.83)	88 (9.40)	106 (11.32)
1985	189 (20.04)	81 (8.59)	88 (9.33)	120 (12.73)
1986	313 (24.49)	120 (9.39)	119 (9.31)	195 (15.26)
1987	343 (20.75)	182 (11.01)	150 (9.07)	291 (17.60)
1988	340 (17.25)	195 (9.89)	331 (16.79)	262 (13.29)
1989	341 (19.01)	184 (10.26)	303 (16.89)	225 (12.54)
1990	433 (19.56)	246 (11.12)	373 (16.85)	265 (11.97)
1991	539 (22.93)	257 (10.93)	338 (14.38)	276 (11.74)
1992	575 (24.04)	274 (11.45)	320 (13.38)	224 (9.36)
1993	537 (24.48)	252 (11.49)	275 (12.53)	177 (8.07)
1994	660 (25.45)	359 (13.84)	216 (8.33)	207 (7.98)
1995	720 (25.86)	298 (10.70)	253 (9.09)	231 (8.30)
1996	696 (24.43)	289 (10.14)	284 (9.97)	261 (9.61)

Cumulative growth rate (% p.a.)

(a) 1979–96	5.73	10.62	8.96	4.14
(b) 1979–88	2.59	15.82	19.62	8.01
(c) 1988–96	9.37	5.04	(–)1.98	(–)0.05

Year	United States	United Kingdom	The Netherlands	Spain
1979	140 (14.42)	28 (2.88)	46 (4.74)	25 (2.57)
1980	140 (11.75)	31 (2.60)	48 (4.03)	21 (1.76)
1981	278 (18.86)	29 (1.97)	37 (2.51)	23 (1.56)
1982	201 (16.92)	24 (2.02)	37 (3.11)	15 (1.26)
1983	209 (20.17)	20 (1.93)	33 (3.19)	19 (1.83)
1984	185 (19.76)	22 (2.35)	27 (2.88)	29 (3.10)
1985	130 (13.78)	25 (2.65)	26 (2.76)	42 (4.45)
1986	111 (8.68)	33 (2.58)	33 (2.58)	45 (3.52)

(Contd.)

Table 6.19 (contd.)

Year	United States	United Kingdom	Netherlands	Spain
1987	120 (7.26)	41 (2.48)	39 (2.36)	63 (3.81)
1988	180 (9.13)	48 (2.44)	42 (2.13)	58 (2.94)
1989	100 (5.57)	49 (2.73)	45 (2.51)	55 (3.07)
1990	87 (3.93)	60 (2.71)	67 (3.03)	86 (3.89)
1991	117 (4.98)	63 (2.68)	72 (3.06)	88 (3.74)
1992	112 (4.68)	68 (2.84)	65 (2.72)	42 (1.76)
1993	121 (5.52)	69 (3.14)	62 (2.83)	38 (1.73)
1994	104 (4.01)	82 (3.16)	59 (2.28)	43 (1.66)
1995	104 (3.74)	84 (3.02)	60 (2.16)	47 (1.69)
1996	121 (4.25)	73 (2.56)	61 (2.14)	50 (1.76)

Cumulative growth rate (% p.a.)

(a) 1979–96	(–) 0.85	5.80	1.67	4.16
(b) 1979–88	2.83	6.17	(–) 1.01	9.80
(c) 1988–96	(–) 4.84	5.38	4.78	(–) 1.84

Note: Figures in brackets represent percentage share of total world export of this fish product.

Source: As in Table 6.1.

COUNTRY ANALYSIS

European and American Countries

Norway

Norway is the largest exporter of dried, salted or smoked form of fish product in the world. The country has been in the business of production and export of this fish product for centuries. She has also modernized the methods of producing these fish items, particularly the fish fillets. But Norway's market share went into a decline in 1981; the loss was considerable during 1988–90, but since then she has started recovering her position, which is also indicated by a considerable improvement in the growth rate of export during the recent period (1988–96).

However, Norway's present thrust area is the export of fish in fresh, frozen or chilled form, whose share in the total export of fish has gone up from 40.68 per cent in 1979 to 71.06 per cent in 1996, whereas the share of fish in smoked, dried or salted form has gone down from 38.13 per cent in 1979 to 20.96 per cent in 1996. These trends indicate that Norway may not like to further increase her share of the world market in the future.

Denmark

Denmark follows Norway in respect of tradition but not in the thrust strategy. Her share of world export in this product has nearly doubled between 1979 and 1996, though of late, there is a decline in her market share as well as her rate of growth in exports of this fish product. The share of this product in the fish basket of Denmark has gone up from 7.35 per cent in 1979 to 13.61 per cent in 1996, whereas that of fish in fresh, frozen or chilled form has gone down from 66.34 per cent in 1979 to 45.45 per cent in 1996. All the indications suggest that Denmark will hold on to her share in the export market of this product in future.

Canada and Iceland

Presently, Canada and Iceland have almost equal percentage shares in world export and both countries are experiencing negative growth rates during the recent period. But for Canada it is a case of increasing her market share by 46.62 per cent between 1979 and 1996, while for Iceland it is a loss of her share by 28.76 per cent during the same period. Canada experienced a spurt in the growth of exports since 1988 as well as in the share of exports, but both began showing a declining trend from 1991. In 1996, the country registered a 10 per cent share of the world's market, which, though small, was still larger than what she had in 1979. The share of this product in the fish export basket of Canada has gone up from 5.97 per cent in 1979 to 12.51 per cent in 1996.

The major thrust areas of Canada appear to be this fish product during the early period and crustaceans in fresh, frozen or chilled form during the recent period, both of whose shares in the total fish exports from Canada have increased whereas the shares of the other two fish products have fallen. But during the past five years, growth in the export of this product has stagnated, and unlike Denmark, Canada will cease to grow in this product market.

In Iceland, the share of this product in the total fish exports of the country has fallen from 27.30 per cent in 1979 to 21.45 per cent in 1996. The country's share of the world export of this product has also fallen. It is likely that Iceland will lose further in this product market.

It is possible that a part of the fish catches of these two countries is being employed for alternative usages. In Canada this may partly be due to a falling trend in catches, but in Iceland it may be due to a deliberate policy choice.

United States

The share of the United States in the world export of this product has sharply decreased by 70.5 per cent between 1979 and 1996; the growth rate has also become negative during the recent period (1988–96). The percentage shares of both import and export are almost the same for the United States. The share of this product in the total fish export basket of the United States has also fallen substantially, from 13 per cent in 1979 to about 4 per cent in 1996.

The United States is a net importer by a small margin of US$ 12 million or 9.02 per cent of total import. There is strong evidence to suggest *prima facie* that the United States is a re-exporter of this fish product, or that the types of fish product that are imported in and exported out of United States are distinct in nature.

All these findings suggest that the exports of this product from the United States will be stabilizing at around the mean exports of 1988–96, that is about US$ 116 million, while for imports, the figure is around US$ 120 million.

United Kingdom, Spain, and The Netherlands

The United Kingdom has steadily increased her exports in this fish product to hold on to her market share more or less consistently. However, the share of this product in the export basket of the country has fallen marginally from 9 per cent in 1979 to about 7 per cent in 1996.

The Netherlands and Spain have lost their market share by 54.85 per cent and 31.52 per cent, respectively, between 1979 and 1996. The Netherlands has improved her growth rate during the recent period, to a positive rate from the negative rate of the earlier period (1979–88), but the share of this product in the total fish exports of the Netherlands has gone down from 9.35 per cent in 1979 to about 4.35 per cent in 1996

Spain's rate of growth in export has become negative during the recent period. The share of export of this product in the total exports of fish from Spain has also fallen from 6.15 per cent in 1979 to about 3.50 per cent in 1996. Both The Netherlands and Spain are likely to experience a further fall in their share of the export market.

The country-specific analysis made above suggests that barring Denmark all the other countries of America and Europe are either moving away from the export of this rather low value fish product or are not willing to increase their market share any further.

TABLE 6.20
Export of fish in salted, dried or smoked form
by major exporting countries of Asia

(in million US dollars)

Year	Hong Kong	Indonesia	China	Singapore
1979	11 (1.13)	5 (0.51)	–	9 (0.93)
1980	15 (1.26)	7 (0.59)	–	9 (0.76)
1981	17 (1.15)	4 (0.27)	–	12 (0.81)
1982	19 (1.60)	6 (0.51)	–	13 (1.09)
1983	19 (1.83)	5 (0.48)	–	9 (0.87)
1984	22 (2.35)	5 (0.53)	–	10 (1.07)
1985	26 (2.76)	4 (0.42)	–	12 (1.27)
1986	26 (2.03)	4 (0.31)	–	15 (1.17)
1987	36 (2.18)	7 (0.42)	–	22 (1.33)
1988	44 (2.23)	14 (0.71)	16 (0.81)	24 (1.22)
1989	44 (2.45)	20 (1.11)	11 (0.61)	22 (1.23)
1990	47 (2.12)	28 (1.27)	94 (4.25)	23 (1.04)
1991	58 (2.47)	34 (1.45)	26 (1.11)	29 (1.23)
1992	63 (2.63)	45 (1.88)	54 (2.26)	44 (1.84)
1993	65 (2.96)	53 (2.42)	63 (2.87)	41 (1.87)
1994	73 (2.82)	84 (3.24)	133 (5.13)	55 (2.12)
1995	93 (3.34)	69 (2.48)	133 (4.78)	57 (2.05)
1996	108 (3.79)	88 (3.09)	76 (2.67)	51 (1.79)

Cumulative growth rate (% p.a.)

(a) 1979–96	14.38	18.38	–	10.74
(b) 1979–88	16.65	12.12	–	11.51
(c) 1988–96	11.88	25.83	28.86	9.88

Note: Figures in brackets represent percentage share of total world export of this fish product.
Source: As in Table 6.1.

Asian Countries

Table 6.20 analyses the export performance of major countries of Asia in this fish product group.

The countries mentioned in Table 6.20 represent 67.30 per cent of Asia's share of export in dried, salted or smoked fish. The remaining countries are Korea (DPR), Korea (R), Thailand, India, Sri Lanka, Japan, Vietnam, Pakistan etc., each having a small share of the world export of this product.

Hong Kong

Hong Kong is the leader in the export of this fish product from Asia. Her share in the world export of this fish product has increased by 235.40 per cent between 1979 and 1996, and the growth rate in her exports has been consistently good. The share of this product in the total export of fish products from Hong Kong has also increased substantially, from 5.65 per cent in 1979 to 22.40 per cent in 1996.

It should be recalled that Hong Kong is a net importer of this product with a margin of US$ 205 million, that represents 65.50 per cent of the total import. Hong Kong is known to be a re-exporting country for many types of fish products, including dried, salted or smoked fishes.

Indonesia and China

These two countries have considerably increased their share of the world export of this product. The share of Indonesia increased by more than six times between 1979 and 1996, and that of China by more than three times between 1988 and 1996. Their rates of growth in exports of this product are also impressive, with Indonesia increasing the rate of growth by more than double during the recent period (1988–96) as compared to what she had achieved in the earlier period (1979–88). Both Indonesia and China, like Norway, traditionally produce these fish items but their entry to the world export market in a big way is a recent phenomenon.

The share of this product in the export basket of Indonesia has gone up from 2.25 per cent in 1979 to 5.25 per cent in 1996. For China, the share has increased from 1 per cent in 1998 to about 2.7 per cent in 1996. Indonesia and China, along with the other Asian countries, are trying to capture those markets that are being left out by Europe.

Singapore

Singapore is also a re-exporting country in this fish product. As of 1996, her export and import exactly match each other. She has also increased her export of this fish product (like all other Asian countries) by 92 per cent between 1979 and 1996. The share of this product in the fish export basket of Singapore has remained the same, at about 9.25 per cent, during the period under study.

Market Competitive Index

Table 6.21 analyses the comparative market expansion in this product by the countries discussed above.

Table 6.21

Comparative expansion of export of dried, salted or smoked fish products by major exporting countries of the world

(per cent)

Country	Market expansion 1979–96	Market expansion 1987–96
Norway	157.78	102.92
Denmark	455.77	58.79
Canada	330.30	89.33
Iceland	99.24	(–) 10.31
United States	(–) 13.57	0.83
United Kingdom	160.71	78.05
The Netherlands	32.61	56.41
Spain	100.00	(–) 20.63
Hong Kong	881.82	200.00
Indonesia	1660.00	1157.14
China	–	375.00
Singapore	466.67	131.82
WORLD	193.41	72.35

The Market Competitive Indexes of major exporters of fish in dried, salted or smoked form are presented in Table 6.22.

Projections

A regression analysis of the export data of all the four groups of fish products for different countries was done. But due to high volatility in the exports of some countries in some groups of fish products, a statistically significant regression equation could not be established in some cases. In

TABLE 6.22

Market Competitive Indexes for major exporting countries of the world in the international trade of dried, salted or smoked fish products

Country	Market Expansion Index (1987–96)	Market share (%) (1996)	Degree of fluctuations (1 – CV)	Comprehensive market parameter (col. 2 x 3 x 4)	Market Competitive Index	Country ranking
(1)	(2)	(3)	(4)	(5)	(6)	(7)
Norway	1.42	0.2443	0.7495	0.2600	31.58	1
Denmark	0.81	0.1014	0.8087	0.0664	8.06	5
Canada	1.23	0.0997	0.8494	0.1042	12.65	3
Iceland	(–) 0.14	0.0961	0.8712	(–) 0.0117	(–) 1.42	12
United States	0.01	0.0425	0.7864	0.0003	0.04	10
United Kingdom	1.08	0.0256	0.8185	0.0226	2.74	7
The Netherlands	0.78	0.0214	0.8441	0.0141	1.71	9
Spain	(–) 0.28	0.0176	0.6906	(–) 0.0034	(–) 0.41	11
Hong Kong	2.76	0.0379	0.6833	0.0715	8.68	4
Indonesia	15.99	0.0309	0.4700	0.0322	28.20	2
China	5.18	0.0267	0.3334	0.0461	2.57	8
Singapore	1.82	0.0179	0.6510	0.0212	2.57	8
Total	–	–	–	0.8235	100.00	

Tables 6.23–6.26 tabulate regression parameters for those countries for whom the trend equations are found to be statistically significant.

TABLE 6.23
Regression parameters of selected countries engaged in the export of fish in fresh, frozen or chilled form

Country	Constant(a)	(b)Coefficient	SE(b)	t. value	R²
Norway	(–) 196.79	120.64	9.02	13.01	0.91
United States	108.93	115.42	9.65	11.96	0.90
Denmark	319.86	46.26	5.25	8.80	0.83
The Netherlands	201.52	36.06	3.46	10.42	0.87
Iceland	196.35	33.24	4.30	7.73	0.79
United Kingdom	36.45	28.51	2.78	10.26	0.87
France	98.66	29.57	3.27	9.04	0.84
Chile	(–) 160.81	43.92	4.33	10.15	0.87
Argentina	4.92	21.99	3.16	6.97	0.75
Indonesia	(–) 110.41	27.50	2.98	9.22	0.84
Thailand	(–) 79.84	26.16	2.15	12.18	0.90
Singapore	10.54	15.39	0.86	17.89	0.95
China	27.11	90.40	13.21	6.84	0.87
Malaysia	(–) 6.71	3.57	0.39	9.13	0.84
Japan	130.16	11.27	1.79	6.29	0.71

Note: At 1% level of significance.

TABLE 6.24
Regression parameters of selected countries engaged in the export of crustaceans, molluscs etc. in fresh, frozen or chilled form

Country	Constant(a)	(b)Coefficient	SE(b)	t. value	R²
United States	70..57	44.85	6.32	7.10	0.76
France	36.74	10.78	1.15	9.39	0.85
Denmark	34.73	19.30	1.79	10.76	0.88
Australia	176.77	23.54	2.48	9.48	0.85

(Contd.)

Table 6.24 (contd.)

Country	Constant(a)	(b)Coefficient	SE(b)	t. value	R^2
Canada	82.85	43.18	4.71	9.17	0.84
New Zealand	32.33	10.09	1.04	9.70	0.85
Argentina	(—) 51.22	19.70	3.17	6.21	0.71
United Kingdom	38.54	18.25	1.66	11.02	0.88
The Netherlands	27.63	11.29	1.10	10.27	0.87
Ecuador	(–) 11.29	36.18	2.38	15.23	0.94
Italy	(–) 1.64	6.09	0.40	15.28	0.94
India	123.70	38.59	4.96	7.78	0.79
Thailand	(–) 424.63	144.38	13.65	10.57	0.87
Indonesia	(–) 60.69	62.15	4.62	13.45	0.92
Bangladesh	6.58	12.51	0.83	15.07	0.93
Philippines	(–) 8.04	18.97	2.41	7.87	0.79
Singapore	(–) 4.20	12.12	1.10	11.00	0.88

Note: At 1% level of significance.

TABLE 6.25
Regression parameters of selected countries engaged in the export of fish in prepared and preserved form

Country	Constant(a)	(b)Coefficient	SE(b)	t. value	R^2
Denmark	63.98	26.95	1.92	14.05	0.93
The Netherlands	2.56	19.46	1.42	13.74	0.92
Chile	(–) 10.50	11.03	0.61	18.04	0.95
United Kingdom	26.73	6.05	0.43	13.98	0.92
France	6.25	7.11	0.75	9.48	0.85
Greenland	27.78	5.05	0.70	7.26	0.77
Thailand	(–) 235.69	100.86	4.91	20.56	0.96
Philippines	12.50	7.47	0.67	11.14	0.89
Malaysia	25.99	4.99	0.52	9.52	0.85
Japan	519.42	(–) 14.84	2.28	(–) 6.52	0.73
Korea (R)	13.61	23.04	1.88	12.27	0.90
Morocco	45.86	7.12	0.80	8.95	0.83

Note: At 1% level of significance.

TABLE 6.26
Regression parameters of selected countries engaged in export of fish in dried, salted and smoked form

Country	Constant(a)	(b)Coefficient	SE(b)	t. value	R²
Norway	150.38	27.30	4.06	6.72	0.74
Denmark	0.13	18.21	1.45	12.52	0.91
United Kingdom	10.96	3.81	0.38	9.99	0.86
Hong Kong	(–) 4.06	5.02	0.40	12.66	0.91
Indonesia	(–) 18.67	4.78	0.64	7.46	0.78
Singapore	(–) 2.33	2.92	0.30	9.68	0.85

Note: At 1% level of significance.

On the basis of the trend equations developed in Tables 6.23–6.26, country-wise projections (non-seasonalized) of export of the four groups of fish products are made in Tables 6.27–6.30.

TABLE 6.27
Projections (non-seasonalized) of export of fish in fresh, frozen or chilled form by selected countries

(in million US dollars)

Country	2000	2001	2002	2003	2004	2005
Norway	2457	2578	2699	2819	2940	3060
United States	2648	2764	2879	2994	3110	3225
Denmark	1338	1384	1430	1476	1523	1569
The Netherlands	995	1031	1067	1103	1139	1175
Iceland	928	961	994	1027	1061	1094
United Kingdom	664	692	721	749	778	806
France	749	779	808	838	868	897
Chile	805	849	893	937	981	1025
Argentina	489	511	533	555	577	599
Indonesia	495	522	550	577	605	632
Thailand	496	522	548	574	600	627
Singapore	349	364	380	395	411	426
China	1202	1293	1383	1474	1564	1654
Malaysia	78	82	86	89	93	96
Japan	378	389	401	412	423	434

Note: Projections are in constant 1996 US $.

TABLE 6.28

Projections (non-seasonalized) of export of crustaceans,
molluscs etc. in fresh, frozen or chilled form by select countries

(in million US dollars)

Country	2000	2001	2002	2003	2004	2005
United States	1057	1102	1147	1192	1237	1282
France	274	285	295	306	317	328
Denmark	459	479	498	517	536	556
Australia	695	718	742	765	789	812
Canada	1033	1076	1119	1162	1206	1249
New Zealand	254	264	274	285	295	305
Argentina	382	402	422	441	461	481
United Kingdom	440	458	477	495	513	531
The Netherlands	276	287	299	310	321	332
Ecuador	785	821	857	893	929	965
Italy	132	138	145	151	157	163
India	973	1011	1050	1088	1127	1166
Thailand	2752	2896	3040	3185	3329	3474
Indonesia	1307	1369	1431	1493	1556	1617
Bangladesh	282	294	307	319	332	344
Philippines	409	428	447	466	485	504
Singapore	262	275	287	299	311	323

Note: Projections are in constant 1996 US $.

TABLE 6.29

Projections (non-seasonalized) of exports of fish in prepared and
preserved form by select countries

(in million US dollars)

Country	2000	2001	2002	2003	2004	2005
Denmark	657	684	711	738	765	792
The Netherlands	431	450	470	489	509	528

(Contd.)

Table 6.29 (contd.)

Country	2000	2001	2002	2003	2004	2005
Chile	232	243	254	265	276	287
United Kingdom	160	166	172	178	184	190
France	163	170	177	184	191	198
Greenland	139	144	149	154	159	165
Thailand	1983	2084	2185	2286	2387	2488
Philippines	177	184	192	199	207	214
Malaysia	136	141	146	151	156	161
Japan	193	178	163	148	134	118
Korea (R)	521	544	567	590	613	636
Morocco	202	210	217	224	231	238

Note: Projections are in constant 1996 US $.

TABLE 6.30

Projections (non-seasonalized) of exports of fish in dried, salted or smoked form by select countries

Country	2000	2001	2002	2003	2004	2005
Norway	751	778	806	833	860	887
Denmark	401	419	437	455	474	492
United Kingdom	95	99	102	106	110	114
Hongkong	106	111	117	122	127	132
Indonesia	87	91	96	101	106	111
Singapore	62	65	68	71	74	76

Note: Projections are in constant 1996 US $.

CONCLUSION

This study has revealed that during the past two decades some distinct changes are taking place in the global fish markets both in respect of consumption behaviour of fish eaters and import–export status of countries in the market place. In some cases, old truths are getting re-established after being in hybernation for some time. In other cases, newer directions of trade are becoming evident. For example, as we have seen in Chapter 5, fish in fresh, chilled and frozen form is still dominating the world fish market, and no let-up in the trend is evident in the future. The same is the case with crustaceans in fresh, chilled and frozen form. America has emerged as the second largest exporter of this product form of fish, and also of crustaceans, due to a domestic cut in consumption and shifting of preference towards the prepared and preserved form. Although the demands of processing technology and quality standards are highest in prepared and preserved form of fish items, it is the Asian countries in the developing region which have emerged as the largest exporters of this fish product to the developed parts of the world. Europe's export and import of fish in fresh form closely match each other, which suggests an element of re-export by European countries. In crustaceans and fish in prepared and preserved form, Europe has become a net importer. Africa is emerging as a new competitor in the world market of both fish and crustacean products in fresh forms.

Though traditional fish trade in salted, dried and smoked form has a low and declining market share, it is very unlikely that it would go out of the world fish market. This product is highly culture specific and hence there still are markets for this product, both in Europe (for example Denmark, Norway), America (for example United States), and Asia (for example Japan).

The changing world market scenario necessitates changes in the export strategy of nations engaged in the fish trade. The Market Competitive Index for different fish products across the world markets will help the countries to prepare competitive strategies for future growth.

II
INDIAN FISHERY

7

Structure of Production and Terms of Trade in the Indian Fishery Industry

INTRODUCTION

India has an Exclusive Economic Zone (EEZ) of size 2.02 million sq. km. There is, however, some difference of opinion about the continental shelf area and the coastline of the country. According to a study by the Indian Council of Agricultural Research (ICAR), the continental shelf has an area of about 0.4 million sq. km.[1] The measurement by Ministry of Agriculture, Government of India suggests the area to be about 0.5 million sq. km.[2] The ICAR study specifies 5600 km as the length of Indian coastline, while a study by the National Council of Applied Economic Research (NCAER) puts it at 4667 km.[3] The Government of India specifies the coast line to be 8041 km, which is much higher than the calculations made by the above two agencies. There is also a difference of opinion about the distribution of coastline and continental shelf between the East and West Coast of India. For example, according to one calculation, the length of the West Coastline is about 3100 km within a continental shelf of 2.05 lakh sq. km[4] whereas the Government of India publication cited above places these at 3473 km

[1] Prasad, Raghu R. and P.R.S. Tampi, 'Marine-Fishing Resources of India', in *Research in Animal Production*, Indian Council of Agricultural Research, New Delhi 1982.

[2] Ministry of Agriculture, Government of India, *Hand Book of Fisheries Statistics*, New Delhi 1996.

[3] National Council of Applied Economic Research, *Export Prospects of Fish Products*, New Delhi 1965.

[4] Joseph, K.M., N. Radhakrishnan, Antony Joseph and K.P. Philip, 'Results of Demersal Fisheries Survey Along East Coast of India', Bulletin of Exploratory Fish Project, 1975.

and 3.57 sq. km, respectively. The report of National Commission on Agriculture, 1976 puts the length of the West Coastline at 3040 km and the area of the continental shelf at 3.94 lakh sq. km. Such wide variations in the measurement of physical areas could be due to different scales adopted by different agencies in a non-Euclidean space or point to point measurement approaches (straight line or circuitous) taken by them. However, there should be some convergence of the techniques adopted by the agencies since the measurement has serious policy implications in regard to developmental investment flows to the coastal regions, which are often decided either on average or on the aggregate.

STRUCTURE OF INDIAN FISHING

Fishing activity in India is a provider of large scale employment. Presently, the fishermen population is about 60 lakh (increased from 9.60 lakh in 1976), of which 40 per cent are engaged on full-time basis, 24 per cent on part-time basis, and the remaining 36 per cent are occasional participants.[5] While fishing does provide employment to a large number of fishermen it does so only at a subsistence level, or at an even lower level. The problem is similar to that of fragmentation of land in agriculture, where each of the fragmented landholding families gets only a small return from its holding. The absence of economies of scale makes the operation inefficient and costly.

The traditional structure of Indian fishing is evident in the types of tools of production used in fishing and their ownership pattern. At present, nearly 67 per cent of fishing crafts engaged in marine fishing in India consist of purely traditional crafts (non-motorized); mechanized vessels are just about 20 per cent of the total boat population; and about 13 per cent are motorized traditional boats. Traditional crafts, including motorized ones, therefore, comprise 80 per cent of the total boat population in India. It is interesting to note here that as early as 1972, India had a total boat population of 230,000, of which traditional fishing crafts (including motorized ones) constituted about 94 per cent and the remaining 6 per cent were mechanized boats.[6] Within a span of nearly 25 years, the number of mechanized boats has increased from about 13,000 to 47,000—an increase by 260 per cent—and the number of traditional crafts has fallen from about

[5] Calculated from *Hand Book of Fisheries Statistics*, Ministry of Agriculture, Fisheries Division, Government of India, New Delhi, 1996.

[6] Ministry of Agriculture and Irrigation, Government of India, *Indian Live-stock Census—1972* (Vol. 1), New Delhi.

217,000 to 191,000—a fall by only 12 per cent.[7] This indicates that the traditional sector of Indian fishing has remained predominantly traditional. The addition of 34,000 boats to the mechanized fleet is not so much by way of replacement of traditional crafts but by new entrants to Indian fishing.

This large section of Indian fishermen is also not able to use many of the modern gears invented around the world because traditional fishing boats are not suitable to adjust to the modern gear-technology. Presently, with about 1.90 lakh traditional boats (including motorized boats) fishing within 50 metres depth of the Indian EEZ, per boat catch cannot but be very low. With the advent of mechanized vessels in Indian marine fishing, per boat yield for traditional crafts is going down further because the mechanized boats (even the trawlers with high catch capacity) virtually operate within 50–60 metre depth, that is in the area where only traditional crafts can operate. This has given rise to conflict and severe tension for the past 25 years between the traditional sector and the mechanized sector, with no resolution in sight.

BOAT BONDING

In 1995, the total boat population of India was about 240,000. Marine catches in the same year were 26.92 lakh tons. Assuming, at a modest level, that the capture of a mechanized boat is on an average fifteen times that of a traditional boat and that the capture of a motorized traditional craft is thrice that of a purely traditional craft, the all-India average annual capture of marine fish by a purely traditional craft cannot be more than 3 tons, which cannot be called an economically sustainable level of catches.[8] This explains why Indian fishermen using traditional crafts, these fishermen numbering about 50 lakh—are in perennial bondage with the moneylenders, merchant financiers, and wholesalers for their off-season sustenance and repairing of boats and nets. Due to this bondage, price realization (net of high interest payment) is also low.

[7] Ministry of Agriculture, Government of India, *Hand Book on Fisheries Statistics*, New Delhi, 1996. See also, John, Thomas, Ajith and M. Shahul Hameed, 'Growth of Fishing Fleet of India After Independence', Seafood Export Journal, September, 1995, pp. 31–3.

[8] We should, however, point out that against this all-India average, catches per boat do vary widely from one region to the other. For example, a preliminary investigation made in Tuticorin region reported annual average catch of 19.9 kgs per non-motorized traditional boat using gill-nets. However, even at this rate of catches, the annual net surplus was found out to be only Rs 5201. (See, Sathiadhas, R., R.E. Benjamin and R. Gurusamy, 'Technological Options in the Traditional Marine Fisheries and Impact of Motorisation on the Economics of Gillnet Fishing along Tuticorin Coast, Tamil Nadu', *Seafood Export Journal*, April, 1991.

Our study[9] has revealed that in about 30 per cent cases advances made to fishermen are adjusted against catches at a predetermined price; in about 52 per cent cases these are adjusted against the price determined at the time of the fish harvest. In the remaining 18 per cent cases, various other modes of adjustments are practised. The percentage distribution of the purposes for obtaining these advances is shown in Table 7.1.

TABLE 7.1
Percentage distribution of purposes for taking advances

Purpose	Per cent of responses
Finance for off-season repair and maintenance of boat and nets	57.50
Working capital for acquisition of nets etc.	17.50
For meeting margin money requirement of FIs/ banks for loan for construction of boat	10.00
For personal/family expenses	15.00
TOTAL	100.00

From Table 7.1, the dominant reasons for taking advances by fishermen are for repair and maintenance and for acquisition of nets etc., which together constitute 85 per cent of the total responses. We may also notice that fishermen are required to take advances (though the percentage of such cases is low) for meeting margin money requirement of banks/FIs for loan for construction of boats. This implies that the uniform rules of banks/FIs to insist on margin money creates hardships among the hard-pressed section of the fishermen and hence, despite institutional finance the bondage with the informal financial sector continues.

POOR BOAT ECONOMICS

The low catch per boat and low net price realization does not allow much surplus to be created at the hands of traditional fishermen, making it unable for them to replace the existing crafts by mechanized boats and usher in commercial fishing practices. Unlike other advanced developing nations of South East Asia, like Korea (R) and Thailand, opportunities for alternative employment are not available in India. This has resulted in an increase in

[9] Based on Questionnaire Survey.

the number of people engaged in a boat and a fishing family's total dependence on a single boat. This has the effect of reducing per capita income from a boat further.

Even in cases where a fisherman's family is able to add another craft to the existing one, mostly by raising loans either from the moneylenders or merchant brokers (in which case prospective boat-catches are often found to be bonded even before it is manufactured) or from banks (to which we shall come later), it has been found that in about 83 per cent cases the new boat is a purely traditional craft and in 17 per cent cases it is a motorized traditional craft.

EXCESS BOAT CAPACITY AND ROLE OF FINANCING AGENCIES

During the later part of the 1970s, commercial banks and cooperative banking institutions came in, with the refinance backing from NABARD, to finance boat manufacturing with the avowed objective of snapping the traditional bondage of fishermen with the moneylenders and other middlemen in the fish business. Their advent and approach to help marginal fishermen was laudable, but it added more crafts to the traditional fleet rather than replacing it with mechanized boats. As a result, there was practically no augmentation in income, rather the small kitty got distributed further. Besides, banks were not willing to finance off-season sustenance of fishermen, including working capital for repairing of boats and nets which continued their dependence on non-institutional finance. Our study of commercial banks' reactions to the problem has revealed that the boat-loans suffered from poor repayment from the beginning. These were not viable lending propositions. According to the bankers, the fishermen showed greater allegiance to the moneylenders and merchant financiers in the matter of repayment, despite commercial banks providing loans to them. Enquires with fishermen revealed that to finance their off-season requirement, they have to depend on local financiers who are 'more helpful and informal in providing money when it is required'. Some fishermen also claimed that the interest charged by these lenders is lower than that of the banks! Though this may be true on the face of it (though factual information on this aspect could not be obtained), what is overlooked perhaps is the fact that these financiers might be more than compensating themselves by bonding the catches.

The principles of market economics suggest that a large number of these 'non-viable' fishing units would have died their natural death in order to bring in an equilibrium between fleet capacity and catches. But this did not happen to the expected scale, as is revealed by the fact that there has been only a 12 per cent fall in the traditional fleet during a period of about 25 years. The reason behind this is that in India there is a wide gap between the theoretical and real subsistence level, and many people continue to live much below the poverty line.

MODERNIZATION AND EMPOWERMENT OF FISHERMEN

The above observations do not in any way make redundant the economic principle of reaching an equilibrium between fleet capacity and marine catches, particularly when the latter is exhibiting a declining trend, which is already making life more miserable for traditional fishermen. It is, however, not possible to allow the marginal fishermen to become extinct in the Indian political economy. This is also not desirable from the point of view of the developmental management of the fishing economy. Indian fishermen know the sea by the lines of their palms. Their knowledge of fish species, their stock and depletion level, the migratory behaviour of different species, and also near accurate prediction of catches in a particular season is fascinating. This knowledge is handed down from one generation to the other. The children of a fishing family start knowing all these in their childhood itself. Unfortunately, we have not come across any systematic study to record the methodologies used by the fishermen and the huge knowledge base about marine fishing at the most practical level. If these fishermen are allowed to become extinct this experiential knowledge will also be lost with them, which is a huge social cost that no country can afford to bear. What is necessary, therefore, is to empower these fishermen with (i) modern tools of production without, at the same time, increasing the existing aggregate tonnage capacity, and (ii) financial freedom in operations.

As indicated earlier, the conflict between the traditional sector and mechanized sector in Indian fishing is a serious problem. The regulatory authorities have attempted to resolve the conflict by prohibiting operations of certain mechanized vessels within 50 metres depth. But this measure eluded observance and application at a practical level. Without delving into the reasons behind this, it can be said that this is the age-old conflict between the feudal structure of production and the capitalist structure, in which the latter shall ultimately win. This is the lesson that is obtained from the history of economic development. Any measure, like the one mentioned above, which attempts to maintain the existing feudal structure of production is not only not able to resolve the conflict but also retards the developmental process. Resolution of conflict can be done only by providing a level playing field, which is possible only when the traditional fishermen are empowered with modern tools of production.

It must be emphasized that if the Indian seafood industry has to move along commercial lines and compete with technologically and managerially advanced fish producing countries of the world, its foundation cannot be allowed to remain structurally feudal. It has also to be developed along commercial lines. This is possible only by empowerment of traditional fishermen so that they can come out of the present feudal structure beset in a 'way of life' philosophy, and become commercialized.

SUGGESTED APPROACH

One reason behind the plight of traditional fishermen is that they are both fish-worker and owners of the tools of production. The concept of this artisan economic unit has long been discarded due to its low level of efficiency. A fisherman has not only to fish—which is his principal activity—he has also to worry about maintenance and repair of boats and the fishing gears. While he is self-sufficient in executing the fishing as he has full control over his ability and experience, for the latter he is perennially dependent upon the outside financing agencies, because he is unable to create enough surplus to do it all by himself owing to using a less efficient traditional boat and gear. This syndrome forces the fishermen into a debt-trap. All over the world, artisans ultimately entered into a debt-trap before becoming bankrupt and uprooted. We, therefore, feel that the long-term approach to resolve the issue is to disassociate the fishermen from the ownership of the tools of fishing, namely craft and gear. As the Indian economy is being deregulated with global integration in view, the operation of market forces will ultimately force such a situation to come up, but that will be at a tremendous social cost and pain. The implementation of an official strategy will make the transition less painful. Considering the issues raised for the commercialization of Indian fishing, we suggest the adoption of the following measures.

(1) Corporatization of fishing crafts

(a) The government should encourage the setting up of private sector companies by providing suitable fiscal incentives for manufacturing/acquiring medium-sized fishing vessels equipped with modern fishing gears for the purpose of hiring them to fishermen on rental basis. These companies will maintain these vessels and their gears and offer them for rent on a full seaworthy and catchworthy basis. The fishermen will be responsible for arranging diesel at their own cost.

(b) The Marine Products Export Development Authority (MPEDA) should make the necessary cost–benefit analysis of such projects to decide on the economic rent, besides establishing its commercial and financial viability. If the economic rent is found to be more than what the fishing community can presently absorb, then the government should provide subsidy to these companies to the extent of the difference between the economic rental and acceptable rent.[10]

During the initial years fixation of rentals should be administered by a Standing Committee in every fishing region/state, comprising representatives from the fishermen community, the vessel owners (companies or otherwise), MPEDA, and the respective state governments.

(c) The subsidy amount should be equally shared between the central government and the respective state governments. There should be separate rules of registration with the Harbour authorities, and other such agencies, for vessels put on hire.

(2) Role of banks and financial institutions

(a) As a matter of policy, banks and financial institutions should stop financing any new traditional fishing craft. Their primary objective should be to provide finance for the replacement of existing traditional crafts by mechanized vessels. Corporatization of fishing crafts will fit into this objective. This, however, should not stop them from financing any individual venture for acquiring a mechanized fishing vessel. Preference should be given to those ventures which offer the vessels on hire to fishermen. Rate of interest chargeable on financing all such vessels, whether owned by a company or individual, should be at par with the financing of agricultural implements.

(b) Under the proposed scheme, the cost of operations for fishermen hiring vessels should be limited to hiring charges payable per trip and the cost of diesel. (It is intended that the vessel owner will provide operators and maintenance crew on board, the cost of which will be included in the hiring charges.)

A fishermen will apply to the bank for setting up an annual loan limit, giving details of the number of trips he intends to make in a given year and the expected catches from such trips. The bank, after being satisfied with the credibility of the applicant in undertaking such fishing operations, will sanction an annual limit covering full operational costs of 90 per cent of such trips. No margin money is proposed on the costs of each such trip, as

[10] However, due to economies of scale the cost of repair and maintenance is expected to be much lower than what are incurred by the private boat owners. This will have the effect of lowering down the economic rent and bringing it closer to the acceptable rent.

it is seen that for providing margin money on various bank loans raised by the fishermen they have to borrow or take advances from moneylenders or other intermediaries. At the time of making a review of the account in the next year, if it is found that the applicant has not undertaken all the trips as per projections, his limit for the next year will be reduced to the extent of 90 per cent of the trips that he undertook in the last year. In cases where the actual trips equalled or exceeded the projected trips in the last year, then the following year's limit will be 90 per cent of the projected trips of the following year.

Hiring charges and the cost of diesel will be directly paid by banks to the vessel owner and the diesel supplier as the case may be.

(c) Port authorities should provide a common space within the Fishing Harbour for banks to open their extension counters under one roof with common facilities, the overhead costs of which will be shared by all the participating banks. Banks can also think of installing a common cash vault with numbered locker-type cages—each bank using one locker-cage—to minimize the cost further.

These extension counters will be responsible for the disbursement of loan, calculated on trip basis within the limit set up by the bank. It is desirable to have the accounts of the vessel owners and diesel suppliers at the different extension counters of the bank to facilitate disbursement of loan to the advantage of all the three parties. Bankers will work in close collaboration with the Marketing Association at each Fishing Harbour. In fact, banks' representatives may also be co-opted as members of the Marketing Association. The extension counters of respective banks will send a statement to the Marketing Association containing the names of individual fishermen and amount of advance made to them. The Marketing Association will be charged with the responsibility of issuing gate passes to buyers of fish from the Fishing Harbour.

The price of the fish settled between the buyers and fishermen at the auction hall will include the cost of trips and diesels on production of the recovery bill issued by banks making advances to the fishermen. This recovery bill will include the principal amount of advances plus an estimated interest amount. It will be the responsibility of the buyer to deposit the hiring and diesel charges incurred by the fishermen with the Marketing Association before obtaining gate passes. At the end of the day the Marketing Association will remit the recovery amount to the respective banks as per the statement received by them from such banks.

The extension counter of the bank will not make further advances if the earlier ones are not repaid in full. The interest payable to banks on such advances will be equal to that payable on agricultural crop loans.

By working out a collaborative arrangement, within or without a legal framework between banks and Marketing Associations of different Fishing Harbours/landing centres, banks will not have much problem for recovery of their advances.

POLITICAL ECONOMY OF TERMS OF TRADE AND PRICE DETERMINATION

We have mentioned earlier that the fish market is predominantly controlled by middlemen in the form of moneylenders and merchant financiers. Advances taken by fishermen on bonding require them to deliver the materials to the financier who may be directly operating as wholesaler or has a linkup with one or more wholesalers (sometime money lenders are also wholesalers). However, owing to organized pressure from fishermens' associations and cooperatives in some large harbours, it has been possible to some extent to separate the lending function and procurement operation in some such harbours. But in other harbours and landing centres the old practice continues. When boats are not bonded, fishermen have a choice to deliver the materials to a wholesaler of his choice, but they would invariably deliver to a wholesaler only and not directly deal with fish traders or exporters, apparently for ease of operation, but mostly under the pressure of the prevalent practice.

Procurements at the landing points are almost totally controlled by wholesalers having substantial money and political power. It is difficult, if not presently impossible, for domestic fish traders and processors-cum-exporters to break the wall and participate in the auctions at the harbours or landing centres directly. The perishability factor plays a major role in determining the price when there is a bumper catch. The financial condition of the fishermen is not such as to enable them to carry the stocks for better price realization. Even if they could, the non-availability of cold storages within or nearby the landing centres would make it difficult to carry the stocks. Cold storage facilities are not available except at major harbours and landing centres. Most of these cold storages are also owned or controlled by traders. Government owned and run in-harbour cold storages are very few, and their operational efficiency is also very low. In the absence of carrying capacity of stock, the fishermen are forced to sell their bumper catches at throwaway prices. It has also been seen at times of bumper catches that fishes of undryable varieties are thrown around the beaches, simply because buyers are not available. A major part of the fishing effort is thus lost. The established practice in the primary market (harbours or landing centres) is that all fish landings must clear out the same day (in fact, within 4–5 hours of landing), whether bought or wasted. There is a vested interest in maintaining the status quo. The nexus is controlled by merchant financiers and wholesalers.

The materials for the fish trade, both domestic and export, are procured from the wholesalers in auctions. For domestic trade, the procurement price is primarily determined by the supply and expected demand of the consumers in seasons. The contact between the trade and the wholesaler is either direct or through distributors (aratdars). In some harbours or landing centres, the wholesalers may not own the materials (though they might have financed the fishermen) but would act as the agent of the fishermen. Materials are delivered to the trade and sold to the consumers in fresh-chilled form. Fish in fresh-frozen form is yet to be acceptable in the domestic market, though this is bacteriologically superior to the fresh-chilled form. The mindset of Indian fish-eaters and the rather higher cost of frozen materials are the two deterrent factors.

Processors-cum-exporters enter into the trade generally through fish brokers, who are mandated to procure a certain quantity of materials within a given price range. The broker is paid an advance to procure fishes on behalf of the principal whose name may or may not be disclosed to the wholesalers. A broker may take part in auctions for more than one principal. For large exporters, the procurement function is performed by specialized buyers in their employment who may engage a fish broker or operate independently. These brokers or the specialized company-buyer play an important role in the export trade of fish products. They set the price trend by providing two-way information. They are also expected to have special knowledge about the quality and size of fishes in a given lot of fish put under auction by the wholesaler. The processor-cum-exporter may find himself landed with a large number of fish of the wrong size or otherwise unexportable quality because of a wrong decision of the buyer engaged by him.

The constituents of the fish export trade in India at present are as follows:

(a) *Processors-cum-exporters:* They are the major constituents of the fish export trade. This is an organized sector with a representative trade association, by the name of Seafood Exporters Association, which acts as a link between the trade, government, and the MPEDA. Nearly 95 per cent of processors-cum-exporters are small-scale industries (SSI), and the remaining about 5 per cent are large houses and seafood export divisions of multinational companies like Hindustan Lever, ITC etc.[11] They fully participate in the value chain, from procurement to value addition through

[11] Editorial, *The Seafood Export Journal,* December 1996, p.3.

different levels of processing. However, for nearly 80 per cent of processors-cum-exporters, the value chain ends at block freezing and/or Individual Quick Freezing (IQF). Truly speaking, it is more value preservation than value addition. Real value addition, which is popularly known in the trade as consumer packs, is done by the top 20 per cent of the trade. But in the absence of reservation, this segment also processes and exports block-frozen or IQF materials which should be the domain of SSI Sector.

(b) *Processors:* These firms originally started as processor-cum-exporters and installed processing capacity in anticipation of exports. But owing to cash flow problems due to losses incurred during the recent crisis they could not undertake exports on their own. In order to make a living and absorb at least a part of the overheads, they offer the processing facilities to merchant exporters or other processors-cum-exporters who need them on payment of processing charges or on lease/rental basis.

(c) *Merchant exporters:* These firms do not have any processing facilities. They procure raw materials from the fish market, get them processed at any processor's factory by paying processing or labour charges, and export them in their own name. They may also buy finished products from other exporters who are unable to find an overseas buyer at a given/expected price.

Merchant exporters generally do not have a long-term stake in the industry. In this trade, both entry and exit are easy. Unlike a processor-cum-exporter, a merchant exporter does not need to have any investment in fixed assets, and can always hire the processing services. What he needs is an export order and money to deliver the consignment. When the going is bad he can simply walk out without incurring any great loss. For many, the attraction to join the fish export trade is not so much for making export, but to enjoy some of the export incentives, like import licenses etc. which have high values in the market. Prior to the liberalization of the economy, when import licenses were not easy to come by, there had been a mushrooming of merchant exporters in the seafood export industry. The other incentive to join the trade is to convert black money into white. This industry is predominantly a cash economy, right from the procurement of raw materials onwards. Even if an export is booked at a loss the underlying purpose is served and, since the export profit is not taxable, the revenue authorities do not pay much attention to this sector.

As a result, when a merchant exporter enters into procurement he distorts the normal process of price determination. His bidding price in an auction is often higher than what it should be by the demand–supply equation. Being invariably a volume buyer, his bidding sets in motion a higher price trend. The processor-cum-exporters lose in the bargain.

Besides the above, fishery products exports also take place through export houses which are popularly known as 'route-through exports'. There are three categories of export houses in India, namely (i) Super Star Trading House; (ii) Star Trading House; and (iii) Export House. They are categorized as such by their volume of exports. The quantum of some of the export incentives varies with the status of an export house for the same volume of export. For example, the import entitlement of a certain value of export for a processor-cum-exporter may be 10 per cent, which may go upto even 30 per cent for a Super Star Trading House. This provides an incentive to these Houses to get exports routed through them. In the bargain, the concerned export house shares a part of the incentive in cash value with the actual exporter. Cash-strapped exporters often join the bargain.

PRICE EQUATION

For procurement of raw materials, the seafood export industry is in the sellers' market, controlled not by the fishermen but by the wholesalers, as mentioned before. During the boom period, unplanned capacities, particularly in block-frozen and IQF, were created. As a consequence, the industry is presently reeling under the burden of capacity overheads. The immediate reaction to this has been a drive towards utilization of capacity and to thus absorb the overhead. This has heightened the pressure on the demand for raw materials. Added to this is the insistence of banks on export turnover equivalent to a minimum of five times the working capital (for some banks it is as high as ten times). These two together have made the industry volume-driven.

The price equation that is presently determining the procurement price of exportable fish materials for an individual exporter is as follows:

Price of raw materials = (estimated export realization + export incentives) − (processing cost + freight & insurance + interest + export cess)

The supply market (wholesalers) has complete knowledge about export incentive, freight & insurance, and cess, and a fair knowledge of the estimated export realization (in block-frozen form) and its processing cost. What is not known to the market is the interest cost of an individual unit and its own estimate of the export price of a unit of fish product. Hence, the market ignores the interest cost and puts in its own estimate of export price realization in determining the price of raw materials. The evolving market norm is such that, on the one hand, exporters are forced to pass over to the market any additional export benefits that could have accrued to them, like devaluation of Indian rupees or lowering down of cess, and, on the other, the market absorbs the cost of revaluation or rise in cess.

Presently, in the above equation export incentive is the only profit element for an exporter. The most important component of export incentive is the duty exemption under the Duty Exemption (Pass Book) Scheme (DEPBS), which is available only after the export is made and the bank routing the export makes an endorsement in the Pass Book. The benefit available under DEPBS may be eaten up if the interest cost of a unit is high or it has made a wrong estimation of the expected export realization. All these make the industry volume-driven, not profit driven. In fact, presently, for the majority of industry members, the more the exports they make, the more are the losses they incur. However, they are forced to make exports for the reasons mentioned above. As a result, even though seafood exports of the country have risen, the industry is becoming sick. Chapter 10 makes a set of recommendations to get the industry out of this vicious circle and make a turnaround.

IMPACT OF DEVELOPMENTAL PLANS

Since independence in 1947, all sectoral developments in India have emanated from successive Five-year Plans. While development of fishery too has found a place in all these plans, it has always received lesser attention that what is due to it.

The First Five-year Plan (1950–1 to 1955–6) did not do much in terms of allocation of funds for fishery development. The allocation was a meagre Rs 5.14 crore, representing only 0.26 per cent of the total outlay of Rs 2013 crore. The actual expenditure was even much smaller, only Rs 2.80 crore. But the plan document brought together in a coordinated manner the thrust areas already existing, though in a disjointed manner, before the plan period commenced. These were as follows:

(a)   Motorization of traditional boats and the introduction of mechanized vessels in Indian fishing;

(b)   Introduction of mothership vessels along with large trawler for offshore operations;

(c)   Development of harbours and landing centres, and provision of adequate supply of ice, construction of cold storages, and transport infrastructure;

(d)   Development of export marketing of Indian fish products;

(e)   Provision of training facilities for fishermen in the use of modern fishing techniques.

The above objectives and thrust areas more or less remained the same during the subsequent plans, with a slight change in emphasis in one objective or the other in different plans.

In the Second Plan (1956–61), an integrated approach was taken to link ultimate utilization of fish to the transportation, storage, and marketing of fish, which required, among other things, development of fishing harbours and landing centres. The thrust on deep-sea fishing and improving fishing methods led to exploratory surveys to locate new fish stocks and the introduction of about 1800 mechanized vessels in Indian marine fishing. The latter encouraged fishing boat building yards in Gujarat, Maharashtra, Karnataka, Kerala, Tamil Nadu, and Andhra Pradesh.

The plan outlay was Rs 12.26 crore for fisheries development, which was about 0.26 per cent of the total plan outlay of Rs 4800 crore during the Second Plan. The actual expenditure was still lower at Rs 9.06 crore. In both the two plans, the percentage allocation of resource outlay remained the same.

The Third Plan (1961–6) laid emphasis on an increase in fish production, which brought together on the one hand schemes for improvement of the conditions of fishermen, imparting training in improved methods of fishing and development of fisheries cooperatives, and on the other, development of export trade. The emphasis on mechanization of Indian fishery and exploration survey for increasing fish production continued in the Third Plan also.

To operationalize the plan objectives, the introduction of about 4000 new mechanized boats was proposed and four new stations for exploratory survey were planned to be established at Veraval, Mangalore, Paradip, and Port Blair.

In order to help boost up fish production and export, the Plan aimed at providing landing and berthing facilities at 16 ports and to establish 72 ice and cold storage plants in different states.

The cooperative movement in Indian fishing led to the establishment of about 2100 fisheries cooperatives with a membership of about 2.2 lakh mainly located in Maharashtra, Gujarat, Andhra Pradesh, Kerala, and Tamil Nadu. But they began showing their weaknesses during the Plan period itself. More than 60 per cent of these cooperatives became sick by the end of the Third Plan, and, within next two Plan periods, the cooperative movement in Indian fishing virtually failed.

The allocation of funds for fisheries development during the Third Plan increased marginally from 0.26 per cent in the earlier two Plans to 0.38 per cent of total outlay of Rs 7500 crore. Out of a total Plan outlay of Rs 28.27 crore, an amount of Rs 23.37 crore was finally spent.

The Third Plan was followed by three Annual Plans during 1966–7 to

1968–9. These three Annual Plans aimed at consolidating the efforts undertaken during the last Plans. No new objectives or thrust areas were proposed. At the end of these three Plans, about 5700 mechanized boats were introduced and work for improving landing and berthing facilities at 30 ports were taken up. In fact, the outlay objectives of the Third Plan were more than achieved by the end of these three Annual Plans. The Plan outlay during these three Plans was Rs 42.21 crore, out of which an amount of Rs 31.67 was actually spent.

In the Fourth Plan (1969–74), an increase in fish production along with improvement of fisherman's economic condition and the development of export potential were emphasized. The Plan document recognized for the first time that the country's demand for fish would exceed supply.

About 5500 mechanized boats and 300 large fishing trawlers were proposed to be introduced during the Plan with financial aid from the World Bank. The government made a provision for suitable subsidy for the acquisition of vessels by entrepreneurs. A survey on deep-sea fishery was launched with the assistance of the United Nations Development Programme and under the Indo-Norwegian Fishery Development Project. Emphasis was laid on providing berthing facilities for larger vessels. Forty-eight minor ports were identified for the development of berthing facilities for small boats. It was also proposed to set up 10 composite plants and 73 ice factories and cold storage plants in different coastal states of the country. From this Plan financial institutions like the Agricultural Refinance Corporation (later, National Bank for Agriculture and Rural Development), Industrial Development Bank of India etc. were involved in financing a part of plan outlays.

However, all the plan targets could not be achieved. Only 2300 mechanized vessels and 76 large trawlers could be introduced as against the targeted number of 5500 and 300 respectively. But other targets were more or less fulfilled. The Plan outlay for fisheries development was also increased in this Plan to 0.52 per cent of the total plan outlay of Rs 15,902 crore. But out of the Rs 82.68 crore so proposed, only Rs 54.11 crore could be utilized, which might have contributed to the reduction of the percentage allocation in the subsequent Plans.

The Fifth Plan (1974–9) recognized that Indian fishery resources were not adequately exploited and there was much scope for further exploitation. Hence, efforts for exploring and exploiting fishing resources were intensified by the introduction of the Trawlers Development Fund through which assistance was provided to acquire vessels from abroad for deep-sea fishing. Some 4000 additional mechanized boats and 200 large deep-sea fishing vessels were proposed to be introduced during the plan period. Steps were also taken for the acquisition of large sophisticated vessels from abroad and also from the Goa shipyard for undertaking exploratory survey work. The State Fisheries Corporations were encouraged to develop diversified fishing, processing, and marketing activities by providing suitable financial incentives. Fish Farmer's Development Agencies were set up to promote intensive aquaculture in selected districts. Pilot projects were introduced in all coastal states for development of brackish water fish farming.

The total outlay for fishery development in the Fifth Plan, including one Annual Plan that immediately followed, was Rs 151.24 crore, which worked out to be about 0.38 per cent of the total plan outlay of Rs 39,322 crore. The actual utilization was Rs 134.98 crore.

The Sixth Plan (1980–5) continued the emphasis on increasing fish production both in the marine and inland sectors, improving the socio-economic conditions of fishermen, and exporting of fish products. These were planned to be achieved by organizing a National Fishery Survey in the Indian EEZ, introduction of additional mechanized vessels in the Marine sector, and the implementation of extension programmes for education and training in modern methods of aquaculture production along with provision of inputs for increasing production. It was proposed to intensify the efforts on processing, storage and transportation of fish, and tap new markets for the exporting of fish products. The investment outlay for fisheries development continued to be 0.38 per cent of total Plan outlay of Rs 97,500 crore. Out of the proposed expenditure of Rs 371.14 crore, only Rs 286.95 crore could be utilized.

The Seventh Plan (1985–90) envisaged to work towards achieving the targets set by the nation to be achieved by the end of the millennium. Wide-ranging proposals were made for overall fisheries development of the country.

Investment in deep-sea fishing was stepped up for exploiting fishery resources beyond 50 metres. A total of 25,000 mechanized boats and 350 deep-sea vessels operating in the Indian EEZ by 1990 was envisaged. The number of fishing harbours/landing centres was to be increased to 140 during the Plan period. It was proposed to strengthen the traditional sector by the introduction of Fiberglass Reinforced Plastic (FRP) boats, upgradation of the boats by motorization, and the introduction of new gears for diversified fishing activity. The growing conflict between the traditional fishermen and the mechanized sector was proposed to be resolved by vigorously implementing the provisions of existing legislations towards this end. Social security measures like insurance, and other welfare schemes were introduced for the benefit of fishermen.

The Plan laid emphasis on bringing in modern post-harvest technology of preserving, processing, and marketing of fish. It proposed to set up hygienic fish markets and an integrated cold chain to facilitate marketing of fish products. The establishment of Fisheries Industrial Estate in selected fishing villages with required infrastructural facilities was also taken up during the Seventh Plan.

For further development of inland fisheries, emphasis was laid on intensive high-yielding fish farming in tanks and ponds through Fish Farmer Development Agencies. Area Development schemes were introduced to establish Prawn Farming Estates, particularly in brackish water areas, through schemes like Integrated Brackish Water Fish Farm Development. The setting up of fish hatcheries was encouraged along with the introduction of in-door hatchery facilities in commercial seed farms. The Plan also proposed to reorganize and strengthen the organizations engaged in fisheries development, such as the Central Institute of Fisheries, Nautical and Engineering Training, Fisheries Survey of India, Central Institute for Coastal Engineering for Fisheries, etc.

However, despite all these laudable objectives, the percentage share of outlay for fisheries development in the Seventh Plan was reduced to 0.30 per cent of the total Plan outlay of Rs 180,000 crore. Amount-wise, it was Rs 546.54 crore, out of which Rs 487.56 crore was actually utilized.

At the end of the Seventh Plan, the number of mechanized vessels and deep-sea vessels had increased to 23,848 and 168, as against the targets of 25,000 and 350, respectively. However, hectarage in aquaculture was increased from 55,000 in 1980 to 592,533 by the end of Seventh Plan.

The two Annual Plans between the Seventh Plan and the introduction of Eighth Five-year Plan in 1992 implemented the unfinished agenda of Seventh Plan. The outlay in these two Annual Plans was Rs 292.74 crore, out of which Rs 272.68 crore was utilized.

The Eighth Plan (1992–7) called for the optimum exploitation of both marine and inland fishery, and conservation of fish and fishery resources. Emphasis was laid on generating direct and indirect employment opportunities in the seafood industry and improving the socio-economic conditions of fishermen by raising their productivity level through training, provision of technical and financial support, and by expanding the scope and coverage of welfare schemes for fishermen, including the construction of fishermen's colonies, provision of proper sanitation, and adequate supply of water. A scheme for providing financial assistance to fishermen to tide over famine conditions during the lean fishing periods was also introduced. In marine fishing, emphasis was laid on enlarging the area of operation of traditional fishermen by motorization of traditional crafts, introduction of off-shore intermediate crafts for exploitation of hitherto unexploited pelagic resources. For deep-sea fishing, gill netting, longlining, etc. were introduced. The motorization of 20,000 traditional crafts and 200 intermediate crafts was proposed.

The development of berthing and landing facilities in major and minor ports continued to receive attention in the Eighth Plan. Two major fishing harbours, seven minor fishing harbours, and ten fish landing centres were proposed to be established during the Eighth Plan. Targets were also set to construct 66 fish handling sheds, an equal number of insulated vehicles for the transportation of fish, and ice plants. Nine modern retail markets were also proposed to be set up during the Eighth Plan.

However, the outlay for fisheries development in the Eighth Plan was only Rs 431 crore, out of a total Plan outlay of Rs 434,100 crore, which is only about 0.10 per cent—the lowest percentage allocation among all the Plans. About 50 per cent of this outlay was meant for capital investments.

The analyses of the contribution of different Plans towards the development of Indian fisheries reveal that the emphasis has almost always been on the development of the marine sector, not so much on inland water fishery, though it is the latter sector which came to the rescue of the seafood industry in India when the marine sector, with its declining trend in catches, failed to cope with the growing demand of fish emanating from the rise in domestic consumption and export imperatives of the country. Except in the Seventh and Eighth Plans, where some attention was given to the development of the inland fishery sector, all the earlier Plans had mentioned it only in passing. As a result, the inland sector has grown without any clear policy direction. The main impetus for its growth has come from the market, more particularly the export market. Indian Plans have not made much contribution towards its growth as compared to the growth of the marine sector, which has been predominantly Plan driven. But the market driven growth of the inland sector has resulted in disasters and created severe environmental problems that have been experienced during the recent period. After the crisis, the country may be entering into an era of controlled development of inland fishery, to which we shall come later.

Another discerning feature is that in none of the plans the investment outlays proposed for fisheries development could be utilized fully. Except in the Fifth and Seventh Plans, (when the utilization level reached its maximum of 90 per cent), the average utilization level had hardly been more than 75 per cent. It is often claimed that the allocation of resources to fisheries development in different Plans is inadequate, but such a claim cannot stand the test of actual utilization of even the 'inadequate amount' allocated. In fact, due to the lower level of utilization the fisheries development entered into the syndrome of lower level of allocation.

TABLE 7.2

Comparison of Plan-period investments outlay and production of fish in India

Plan-period ending	Actual planned expenditure (Rs in lakh)	Fish production (in thousand tons)			Incremental output–capital ratio (col.5/col.2)
		Marine	Inland	Total	
(1)	(2)	(3)	(4)	(5)	(6)
Pre-plan (1950–51)	–	535 (70.95)	219 (29.05)	754	–
I (1955–6)	278	595 (70.83)	245 (29.17)	840	3.02
II (1960–1)	906	880 (75.86)	280 (24.14)	1160	1.28 (4.17)
III (1965–6)	2337	824 (61.86)	508 (38.14)	1332	0.57 (1.47)
Three Annual Plans ending 1968–9	3167	904 (59.24)	622 (40.76)	1526	0.48 (0.65)
IV (1973–4)	5411	1210 (61.80)	748 (38.20)	1958	0.36 (0.62)
V (1979–80)*	13,498	1492 (63.76)	848 (36.24)	2340	0.17 (0.43)
VI (1984–5)	28,695	1698 (60.60)	1104 (39.40)	2802	0.10 (0.21)
VII (1989–90)	48,756	2158 (60.60)	1403 (39.40)	3561	0.07 (0.12)
VIII (1996–97)	40,000**	2513 (51.36)	2380+ (48.64)	4893	0.12 (0.10)

*   Including one Annual Plan
**   Planned outlay
\+   Estimated figure

Note: 1. Figures in brackets in col. 3 and col. 4 denote percentage to total production.
       2. Figures in brackets in col. 6 indicates the ratios calculated with one period lag.

Table 7.2 presents fish production data of both marine and inland sectors and shows it to conform closely the Plan period investments.

As mentioned before, the flow of planned investments was not uniform to the marine and inland sector, rather it was lop-sided towards the marine sector, but it is the inland fishery which has grown much faster than the marine fishery. The cumulative rate of growth of marine fish production from the pre-plan period to the end of the Eighth Plan is 16.73 per cent per annum. During the same period, Inland fish production has grown at a rate of 26.94 per cent per annum.

The inland sector, which had a share of less than 30 per cent of total fish production in the pre-Plan period, has increased its share consistently throughout the subsequent periods and, by the end of Eighth Plan, it now claims a share of nearly 50 per cent of the total fish production of the country. But unfortunately, its rate of growth is showing signs of deceleration, as we shall see later, due to the lack of concerted policy initiatives and environmental problems.

Column 6 of Table 7.2 presents the incremental output–capital ratio of planned investments and fish production in the country. The ratios indicate fish production in thousand tons per Rs 100,000 of planned investment. The ratio is found to be constantly on the decline, and by the end of the Eighth Plan, it has come down to as low as 0.10. Even when this ratio is calculated with a one period lag, the situation does not improve much.

CONCLUSION

The Indian fishing industry is controlled by middlemen in the form of moneylenders, merchant financiers, wholesale traders, and fish distributors. They are the invincible wall between fishermen who make catches and the fish trade, both domestic and export, who reach the fish to the final consumers with or without processing. Although the fish are mostly sold in auction, which is apparently a transparent market mechanism, the net price realization for fishermen's catches continues to be low due to high interest recovery and tendency of the auctioneers to pull down the price. The fishermen are tied to financiers by bonding, not so much for financing their fishing efforts but for off-season maintenance and repair of crafts and gears for which banks would not ordinarily finance. The solution to the problem lies in separating the fishermen's function (fishing) from the ownership of the tools of production (crafts, nets, and gears). A corporate system of ownership and management has been proposed for this purpose.

The procurement price function that has emerged during the recent period is highly adverse to the fish export industry, owing to the penchant of the exporters to utilize their installed capacities and the insistence of the banks to register turnover of not less than five to six times the working capital loan. For the majority of the firms, the more is the export, the more is the loss. But they cannot also stop exporting. What is seen, therefore, is a rising fishery exports for the country at the cost of erosion of net worth of individual units owing to continuous losses. The problem is discussed further in Chapter 10 with suggested solutions.

The Five Year Plans in India—the source of developmental funds flow to different sectors of the economy—have emphasized more on the development of marine fisheries than aquaculture. The unbridled expansion of marine fishing power due to large trawlers and non-enforcement of the limit to their fishing area has not only created overfishing and environmental pollution in the coastal waters, but also resulted in unending conflict between small boat-owners and trawlers operators. This is discussed further in Chapter 8.

With the declining trend in marine catches, India's fish export trade could be sustained only by aquaculture production. This sector has been successful in doing this and presently, aquaculture production commands a 50 per cent share of the total fish production in India. Despite this, the plan allocation and planned approach to its development was minimal. This has led to an uncontrolled expansion of aquaculture, which has created several environmental problems. The need of the hour, therefore, is to pursue sustainable aquaculture development within the appropriate environmental parameters.

8

Marine and Inland Fisheries
of India

INTRODUCTION

The East Coastline of India is about 32 per cent longer than the West Coastline. But on several aspects of fishing it is the East Coast which lags far behind the West Coast. This also reflects the general state of economic development of the states that comprise the two coastlines. The states outlining the East Coast are West Bengal, Andaman & Nicobar Islands, Orissa, Andhra Pradesh, Tamil Nadu, and Pondicherry, whose state of economic development (except Tamil Nadu) is far lower than that of the states of the West Coast, namely Gujarat, Maharashtra, Karnataka, Kerala, and Goa.

With regard to marine fishery, one indicator of development is the percentage distribution of different types of boat population between the East Coast and the West Coast, as given in Table 8.1

TABLE 8.1

Percentage distribution of boat population by types between two coastal states of India

(average of years 1992–7)

Coast/state	Purely traditional boats	Motorized traditional boats	Mechanized boats	Total
EAST COAST				
West Bengal	1.70 (2.55)	0.10 (0.75)	0.80 (4.04)	2.60
Orissa	3.30 (4.93)	1.05 (7.90	0.70 (3.54)	5.05

(Contd.)

Table 8.1 (contd.)

Coast/state	Purely traditional boats	Motorized traditional boats	Mechanized boats	Total
Andaman & Nicobar Islands	0.50 (0.75)	0.10 (0.75)	0.10 (0.50)	0.70
Andhra Pradesh	22.70 (33.93)	1.35 (10.15)	3.80 (19.20)	27.85
Tamil Nadu	11.25 (16.82)	2.25 (16.92)	3.45 (17.42)	16.95
Pondicherry	2.45 (3.66)	0.15 (1.13)	0.25 (1.26)	2.85
Sub-total	42.90 (62.64)	5.00 (37.60)	9.10 (45.96)	56.00
WEST COAST				
Gujarat	3.50 (5.23)	1.80 (13.54)	3.50 (17.68)	8.80
Maharashtra	4.10 (6.13)	0.10 (0.75)	3.35 (16.92)	7.55
Goa	0.40 (0.60)	0.40 (3.00)	0.35 (1.77)	1.15
Lakshadweep	0.30 (0.45)	0.10 (0.75)	0.20 (1.00)	0.60
Karnataka	5.00 (7.45)	0.50 (3.76)	1.50 (7.57)	7.00
Kerala	11.70 (17.50)	5.40 (40.60)	1.80 (9.10)	18.90
Sub-total	25.00 (37.36)	8.30 (62.40)	10.70 (54.04)	44.00 •
TOTAL	66.90 (100.00)	13.30 (100.00)	19.80 (100.00)	100.00

Note: Figures in brackets denote percentage of total of each type of boat.
Source: Handbook of Fisheries Statistics, Ministry of Agriculture, Fisheries Division, Govt. of India, New Delhi, 1996.

From Table 8.1 it is seen that the East Coast has a larger share of boat population in India than the West Coast. It also accounts for a larger share of purely traditional crafts. But in respect of motorized traditional crafts, the East Coast lags far behind the West Coast. In the mechanized sector, the difference between the West Coast and the East Coast is about ten percentage points.

CATCH CHARACTERISTICS

Despite the East Coast having a longer coastline and larger share of the boat population it accounts for only about 30 per cent of the total marine catches. The situation has not changed much since 1978 when the proportion of catches between the East Coast and the West Coast was 28 per cent and 72 per cent, respectively. With such little catch and so high a boat population, particularly in the traditional sector, capture per traditional craft must be much lower in the East Coast than the West Coast and all-India averages. Using the same calculation as in Chapter 7, per-boat catch in the purely traditional sector is found to be about 2 tons in the East Coast, while for the West Coast it is about 4 tons, against an all-India average of 3 tons. It appears that the traditional fishermen of the Eastern Coastal states are far more impoverished than their counterpart in the Western Coastal states. The East coast has too small a piece of the cake to be shared by too high a boat population. This calls for policy attention towards a reduction of boat capacity in the region.

Besides the wide difference in catch quantity between the two coasts, a clear distinction is also observed in terms of the variety of fish species caught in these two coasts. While only ten major fish species account for more than 75 per cent of the total fish catch in the West Coast, a great variety of fishes, numbering not less than 40 species—each accounting for a small quantity—are caught in the East Coast. A large number of shoaling fishes, like oil-sardines, mackerel, or croackers, are found in abundance in the West Coast, but in the East Coast they are few and far between. The reason behind the characteristic differences between the two coasts may be found in the topographic and oceanographic differences between the two coasts. The reason attributed to high fish production in the Western Coast is the continuous upwelling of the Arabian Sea which provides essential food for fish.

PROBLEM OF OVERFISHING

The characteristics of the Arabian Sea appear to be similar to those of the Atlantic and Pacific Oceans, though the latter are in a temperate region while the former, being a part of Indian Ocean, is in a tropical region. It has been found that in temperate oceans, the number of fish species is few but their availability is high by volume. As the varieties are limited, vulnerability of biological equilibrium and amenability to overfishing are very high. People living in such coastal regions has only a fewer variety to choose from. Their tastes also are shaped around these few varieties. Hence, when the demand for fish rises, due to a growth in population and/or rise in income, the pressure on these few varieties increases, which first causes economic overfishing and then ultimately degenerates into biological overfishing. This has happened in the whole of the Atlantic and a large part of the Pacific Ocean. For the same reason this is also happening in the Arabian Sea, though it belongs to the tropical region.

As against the above, when an oceanic region is 'infested' with a large variety of species, not only is the stability in their biological equilibrium comparatively high, they also provide a variety of fishes to the people living in the coastal region. This makes them less dependent on a particular type of fish. As a result, the rising demand for fish is evenly distributed among a large variety. Consequently, chances of overfishing are much less in such an oceanic region.

As in the Atlantic, signs of overfishing are also evident in the Arabian Sea. For example, during 1991–5, catches from the west coast, belonging to the Arabian Sea, have increased by only 8 per cent, while in the east coast near the Bay of Bengal the increase is about 28 per cent. The cumulative growth rates in catches during this period are 1.95 per cent per annum from the west coast and 6.30 per cent per annum from the east coast.

Our survey has revealed that the average size of the individual fish caught in the marine sector is getting reduced, predominantly in the West Coast, during the past three years. The results of the survey are given in Table 8.2.

TABLE 8.2
Survey results on the size of individual fish caught in the marine sector

(*percentage of responses*)

Coast / State	Yes	No	Not responded
East coast	11.31	13.26	1.88
West coast	60.36	5.65	7.54
TOTAL	71.67	18.91	9.42

It may be observed that respondents from the states of west coast are predominantly reporting a fall in the average size of individual fish caught in the marine sector. Except for some of the respondents from Tamil Nadu and Pondicherry, no other state in the east coast is reporting any such fall in size. It appears that fishes in the west coast are not allowed to mature before they are caught. This is a significant indication of overfishing which ultimately results into depletion of stock. In this respect also, the west coast near the Indian Ocean (Arabian Sea) is similar to the Atlantic Ocean. It is of utmost importance now to adopt severe conservation measures, including banning of fishing during the breeding and maturing seasons, to prevent further depletion of fish stocks.

FISHERY RESOURCES OF INDIA

PELAGIC FISHERIES

Pelagic fishes, like Indian oil sardines, mackerel, other clupeoids, tunas, billfishes, flying fish, etc. occupy a major place in the Indian marine fishery. These are also found mostly in the West Coast, though they also show up in some states of the East Coast. The composition of pelagic fishes in the total catch of the East Coast is much lower than that of the West Coast. For example, while in the coastal states of Goa and Kerala pelagic fishes account for more than 80 per cent of total catches, in West Bengal and Orissa the share is less than 50 per cent.

Sardines

Sardines are said to be the tropical counterparts of the herrings of the temperate regions. There are a wide variety of small fishes in this group, of which the Indian oil sardine is most important. About 35 per cent of oil sardines are captured in the West Coast, of which Kerala accounts for 18 per cent and Karnataka, 12 per cent. Of the 65 per cent of oil sardines captured in the East Coast, Tamil Nadu's share is 50 per cent, which constitutes about 77 per cent of all such catches in the East Coast. Like herrings, this fish is also characterized by wide fluctuations in catches.

The production of Indian oil sardines had gone up as high as 302,000 tons in 1968, accounting for more than 33 per cent of the total marine landings in India, and as low as 102,000 metric tons in 1994, accounting for 3.75 per cent of the total. In 1997, the catches almost equalled that of 1972. The extreme fluctuation in catches (with Coefficient of Variation (CV) equal to 42.61 per cent during 1987–97) has made its commercial exploitation for the export market difficult. Hence, it is primarily confined to domestic consumption. Besides reduction in oil, a small part is canned, and in the years of abundance a major part is used as manure.

But if uncertainty is a deterrent to the full commercial exploitation of this fish, it is the very same uncertainty that should be converted into an opportunity by an alternative marketing strategy. If herrings could be made exportable, why can the same not be done for sardines, which belong to the same family.

The mean catch of oil sardines during 1987–97 is about 156,000 tons, with a standard deviation of 66,500 tons. The Working Group[1] has estimated the potential at 191,000 tons. Even if we take the one full standard

[1] Report of the Working Group on Revalidation of the Potential Marine Fisheries Resources of EEZ of India, Marine Products Export Development Authority, Cochin, 1990.

deviation in our estimate, the total catches can be said to vary between 222,000 tons and 89,500 tons, which is nowhere near the potential yield estimate made by the Working Group. These parameters only indicate the highly fluctuating nature of catches, not so much the potential.

If we follow the catch data from 1950 onwards, we might find that there is a pattern, though somewhat loose, even in these fluctuations. A five-to-six year period of low catches (during which the catch may go down to an abysmally low figure) is generally followed by a six-to-seven year period of high catches. This indicates the migratory behaviour of oil sardines. This finding also suggests that despite wide fluctuations in catches, planned commercial exploitation of oil sardines is possible.

Other Sardines, Anchovies, and Clupeoids

Besides oil sardines there are other sardines and clupeoids, which together account for half (55 per cent) of this group of fishes. This inter-catch relationship has remained almost the same during the past 20 years.

The other sardines and anchovies are captured mostly along the West Coast which claims a share of about 63 per cent of the total catches, of which Maharashtra alone has a share of about 50 per cent, followed by Karnataka (13 per cent). In the East Coast which has a share of 37 per cent of the total capture, Tamil Nadu accounts for more than 20 per cent. All the other states of East Coast, except West Bengal, have a share in the total catches of anchovies.

In the case of other clupeoids, like wolf herrings, hilsa, and other shads, etc., the West Coast has a share of more than 70 per cent of the total catches. Kerala and Gujarat share about 27 per cent and 24 per cent, respectively. Of the 30 per cent captured along the East Coast, Orissa, Andhra Pradesh, and Andaman & Nicobar Islands share about 9 per cent, 7 per cent, and 7 per cent, respectively.

The overall fluctuation in the clupeoids group, except for oil sardines, is at a much lower level (CV = 15.67 per cent). Other sardines are fluctuating around a mean of about 84,000 tons with a low CV. The Working Group's estimate of the potential yield is 96,000 tons, in terms of which there is still scope for further exploitation of this species, though at times the actual catches have exceeded the estimated potential.

During 1994–7, the production of anchovies has become stagnant at an average of 36,000 tons from an average of 66,500 tons during 1987–93. The Working Group's estimate of the potential yield for anchovies is 53,000 tons. Though it might seem that the recent stagnancy in the catches of anchovies is due to overfishing much beyond the estimated yield, a close examination of the catch data for the past 25 years reveals that such a low level stagnancy in catches occurred earlier also but the species have turned around vigorously after that. There may be a pattern in the fluctuation of catches which can be exposed fully by a scrutiny of data for a sufficiently long period.

The production of other clupeoids increased considerably since 1991. From an average of 66,500 tons during 1987–90 the catches have increased to an average of about 107,000 tons during the next seven years. The primary impetus for such a spurt in production might have come from the growing export market for this fish product. The Working Group's estimate of the potential yield is 210,000 tons, of which 14,000 tons are coming from a depth beyond 50 metres. In terms of this estimate, even half of this potential is yet to be exploited. However, it may also be that the Working Group might have overestimated the potential. Even during the period of rising growth catches have never exceeded beyond 153,000 tons, and this too has occurred only once.

Catches of other sardines are fluctuating around a mean of 84,000 tons with a low CV (18.36 per cent). The Working Group's estimate of the potential yield is 96,000 tons. There is still scope for further exploitation of this fishery, though at times, the actual catches have exceeded the estimated potential.

The clupeoids group as a whole, including oil sardines, presently account for about 15 per cent of the total marine landings in India. However, the trend suggests that it can go as high as 20 per cent.

Indian Mackerel

The Indian mackerel is the other shoaling fish which is found in abundance in the Indian coastal areas, particularly in the West Coast. Eighty five per cent of the capture is made in this coast, of which Kerala has a share of about 67 per cent, followed by Karnataka (17 per cent); Maharashtra has a small share (about 1.5 per cent); while other states along the West Coast do not have any capture of Indian mackerel. The East Coast accounts for 15 per cent of the total capture of this pelagic fish, of which Tamil Nadu has the largest share (7.5 per cent). The other states of East coast, except Pondicherry, have a small share each.

Similar to oil sardines, the mackerel's wide fluctuation in catches (CV equal to 38.77 per cent) apparently makes it difficult for commercial exploitation. The Indian mackerel also exhibits migratory behaviour, which results in a fluctuation in catches, but the variation is lower than for Indian oil sardines. A CV between 30 and 40 per cent is not an uncommon feature in pelagic fishes. We have also examined the catch data of the Indian mackerel since 1950 to find out whether there exists any regularity in the occurrence of bumper catches like Indian oil sardines. No such phenomenon is observed.

The estimated potential of mackerels is 224,000 tons, of which 62,000 tons are beyond 50 metres depth. If we go by this estimate, there is yet much scope for exploitation of this fishery.

A wide market for this fish exists in the oil rich African countries like Nigeria, Saudi Arabia, Ivory Coast, etc. Nigeria's import of mackerel constitutes about 70 per cent of her total import of fish. The Ivory Coast is the second largest importer in West Africa, after Nigeria. The major suppliers of mackerel to the African countries are the East European countries, dominated by the Russian Federation, followed by Norway, the Netherlands, Ireland, and the United Kingdom. India does not have much place in the mackerel markets of Africa.

Carangids

Besides Indian mackerels, there are other mackerels and carangids like horse mackerel, scads, leatherjackets etc. which contribute between 4–5 per cent to marine fish landings in India.

It has been found that while the horse mackerel is showing some stability in catches, like the Indian mackerel, other mackerels and carangids exhibit heavy fluctuations. The production of total carangids has gone up from 25,997 tons in 1987 to 135,470 tons in 1997, a rise of about 421 per cent. The fluctuations are very high as compared to other fish groups (CV equal to 53.68 per cent with a mean of 104,495 tons). The Working Group's estimate of the yield potential for carangids is 447,000 tons, out of which as much as 304,000 tons are coming from a depth beyond 50 metres. Recent catch data indicate that despite the potential, offshore resources are not being exploited, and catches continue to be confined within the depth of 50 metres.

Tunas and Billfishes

Tunas and billfishes are the other important types of pelagic fishery, not so much by quantity of catches but by their value in the export market. India has a great potential in tuna fishing. It can change the contours of Indian fish exports. Some elaborate discussions are, therefore, necessary. There are a great variety of tunas, by size, genre, and species, which are distributed extensively between two coasts of India. It is interesting to note that though the Pacific Ocean is the largest source of tuna, accounting for nearly 70 per cent of the total world production of tuna, the Indian Ocean follows immediately thereafter with a share of about 17 per cent; the rest is captured in the Atlantic Ocean.

Due to overfishing, several important varieties of tuna, like bluefin, bigeye tuna, are on the verge of near extinction. While the Atlantic and Pacific Oceans have suffered mostly from overfishing, the Indian Ocean is comparatively less exploited.

In the Indian Ocean, skipjack tuna accounts for about 42 per cent and yellowfin for 31 per cent of the total tuna catches. The western Indian Ocean accounts for 80 per cent of tuna catches from the Indian Ocean. But most of the tuna fishing is done by countries like Japan, Korea(R), Spain, and France. India lags far behind.

The eastern Indian Ocean, which contributes 20 per cent of tuna catches from the Indian Ocean, does not produce much of high value tunas. Small-sized tunas constitute about 66 per cent of total tuna catches. Skipjack accounts for only about 11.5 per cent and yellowfin for 8.5 per cent of total tuna catches. Bluefin, bigeye, and albacore together constitute the remaining 14 per cent. During the recent years, except the small-sized tunas, all the other species are showing a declining trend in their catches. The small-sized tunas, captured mostly by the developing nations of the world from the coastal areas, has created some glut in the world market during the recent years, causing aberrations and a general dampening of price. The terms of trade have moved adversely against the developing nations. This calls for alternative policy initiatives by developing nations.

Presently, India has a share of only about 6 per cent of the total tuna catches of the Indian Ocean. The primary reasons behind such a low share are shrimp orientation of the seafood industry of India and the reluctance of catchers, including even large trawlers, to go beyond the coastal waters.

During recent times, tuna catches in India have varied between 36,000 tons and 47,000 tons, representing, on an average, 1.55 per cent of total marine landings. The bulk of these catches occur in the West Coast, which accounts for almost 85 per cent of the total catches. The South-west Coast has the largest share (55 per cent), followed by North-west Coast (30 per cent), while in the East Coast, which has a total share of about 15 per cent, the major part of the capture occurs in the South-east Coast (14.5 per cent).

Tuna landings are known to suffer from wide fluctuations, primarily because of the migratory habit of the species. But in India such wide fluctuations are not noticed owing to the low level of exploitation. The Coefficient of Variation is 17.56 per cent around a mean of 37,573 tons.

The tuna stocks in the coastal area have reached a nearly full exploitation level. Any further expansion of tuna fishery must, therefore, be offshore.

The bulk of tuna landings continues to be from coastal waters, and are low value products. Till the early 1980s, the existence of tuna resources in the oceanic region was unknown to India. It is claimed that India was misled by the major tuna fishing countries in the Indian ocean to believe that tuna stocks were available only in the areas between latitude 4°N and 4°S of the Equator.[2] However, with the Government of India taking the initiative in exploring tuna resources in the 1980s through the Fishery Survey of India, the existence of rich oceanic tuna resources came to be known to Indian fishermen, and by 1991 India's capture of oceanic tunas reached 10,000 tons.

The Working Group's estimate points to the existence of an oceanic tuna resource potential of 209,000 tons, of which the oceanic Exclusive Economic Zone of Andaman and Nicobar Islands alone has a potential of about 1 lakh tons, followed by 50,000 tons in Lakshadweep Islands. But unfortunately, exploitation of these resources is very low, particularly in Andaman and Nicobar Islands where annual catches are only around 5000 tons. As compared to that, Lakshadweep has a better catch performance in tunas. During the past five years (1994–8), the average catches are around 10,500 tons from this region.

Except for Lakshadweep, which is also the traditional bastion of tuna fish, there is not much of a domestic market in this fish. The tuna fish are mostly exported in frozen form to the neighbouring countries, such as Thailand, Singapore, and Malaysia. The price realization is much lower than the world market price of this fish product.

Other pelagic fishes of some significance are the billfish and the flying fish. While billfishes, like the tunas, have a number of varieties and, along with tunas, are widely distributed in the Indian Ocean, flying fishes have a few varieties and are localized in the coastal waters south of Chennai. Both fishes are showing a declining trend, with a total catch of 3147 tons in 1997. India does not have much potential for these two fisheries.

The contribution of the major pelagic fishes discussed here to total marine landings in India was nearly 28 per cent in 1997.

DEMERSAL FISHERIES

In contrast to pelagic fisheries, the demersal fisheries constitute a large variety of fishes. Most important among them are prawns, bombay duck, elasmobranchs, mullets, cephalopoda, eels, and flatfishes like halibut, flounders and soles.[3]

---

[2] Somvanshi V.S., Keynote address to Round Table Conference of Tuna Fishing held at Visakhapatanam on 14 June 1999.

[3] There are some differences of opinion in categorizing fishes under pelagic or demersal category. For example, Prasad and Tampi put ribbon fish and Bombay duck under demersal category, while Joseph puts them under pelagic category (see R. Raghu Prasad and P.R.S. Tampi, 'Marine Fishery Resources of India', in *Research in Animal Production*, Indian Council

Prawns

This is the most important item of sea-catches in India, both by volume and by commercial importance. There are rich fishing grounds along the entire West Coast, while in the East Coast, Tamil Nadu, Andhra Pradesh, and Orissa account for most of the catches. The West Coast, which had a share of nearly 85 per cent of the total prawn landings in 1979, now has a share of about 80 per cent, which in other words means that prawn fishery in the East Coast has expanded, though marginally, during the past twenty years.

Till the mid-1960s, prawns were not considered as a separate fishery. It was only a minor item of the total marine fish captures. No separate statistics were also available as these were grouped under the general head of crustaceans. During the 1970s, the importance of prawns as a high-value export item gained ascendance, and, over a period of time, prawns and shrimps began dominating the seafood export from India. The situation continues till today. Prawns and other crustacean fisheries got a tremendous boost by the recommendation of National Commission on Agriculture in 1976. It said, among other things, that special attention should be paid to crustacean fisheries, which have a great export potential, by a comprehensive survey of the resources, regular monitoring of the status of the fisheries, and diversification of (a) production and processing centres, (b) export products, and (c) export markets.

The above policy direction made India extend her prawn fisheries to newer and newer areas by extensive search operations. Soon the country emerged as the leading exporter of crustacean products. But, at the same time, this policy emphasis on prawns and other crustaceans has made the Indian seafood industry almost wholly prawn and shrimp oriented. It did not diversify itself to other fishery products. There is a general belief that domestic consumers in India cannot absorb the price realization of this product in the export market, and hence virtually no attempt was made to develop the domestic market for this product. The belief is held steadfastly despite the fast growing middle class population in India since the 1980s, who have much more spendable income now than before. The growth of the fast moving consumer goods industry in India and the entry of a large number of multinationals in the field are pointers to the direction. It is time for Indian seafood industry to come out of the decades-old mindset and develop the domestic market for these types of seafood. The existence of a

of Agricultural Research, New Delhi, 1982, and Joseph, K.M., 'Marine Fisheries Resources of India', in *A System's Framework of the Marine Foods Industry in India*, Concept Publishing Company, New Delhi, 1985.

domestic market is always beneficial for a highly export oriented industry, particularly in fish products, as it can absorb the shocks of fluctuation in catches and also the vagaries of the world market.

Presently, prawns account for about 11.5 per cent of the total marine landings in India and, among demersal fishery, claim a share of more than 25 per cent. The penaeid variety of prawn is generally larger in size than the non-penaeid variety and have greater value in the export market. Kerala alone has the largest share of penaeid prawn catches(more than 33 per cent), though Maharashtra and Gujarat together account for the largest share in total prawn catches in India. The highly developed fish economies of these three states are due primarily to the abundance of prawns, the export of which has brought valuable foreign exchange to these regions. These three states together account for nearly 80 per cent of total prawn landings in India.

But the share of prawn in the total marine catches is on the decline during the past ten years. This is primarily due to non-penaeid prawns reaching a stagnation in catches, with a cumulative growth rate of (–)0.96 per cent per annum. The Working Group's estimated potential of 54,000 tons of non-penaeid prawns has already been exceeded a long time back. As against this, penaeid prawns has grown at a compounded growth rate of 5.71 per cent per annum. It must, however, be noted that the present rate of growth in the catches of penaeid prawns is much lower than that of the 12.68 per cent per annum growth rate registered during 1965–75. Although penaeid prawns are yet to reach a saturation point, despite the Working Group's estimated potential of 178,000 tons, there is a general decline in the growth rate of catches of penaeid prawns. The high pressure on prawn fisheries due to exports is causing overfishing.

The mean catches of penaeid prawns during the past ten years was 166,914 tons, with a standard deviation (SD) of 32,034 tons and Coefficient of Variation (CV) of 19.20 per cent. The mean catches of non-penaeid prawns is 98,380 tons, with a SD of 9263 tons and a CV of 9.42 per cent, while for aggregate prawn catches, the figures are 265,294 tons, 35,330 tons, and 13.32 per cent, respectively. These parameters suggest that fluctuation in catches in penaeid prawns is higher than in non-penaeid prawns (low level of fluctuation in the catches of the latter group is due to stagnation in catches) and the overall fluctuation in total prawn catches is much lower than that of tunas.

Other Crustaceans

Besides prawns, lobsters are also high value crustaceans in the export market. Lobsters are found mostly in the rocky regions of Kerala, Tamil

Nadu, and Andhra Pradesh. However, there is now an extension of lobsters fishery in the Maharashtra and Saurashtra coasts.

Lobsters production in India has been falling continuously during the past ten years; the compounded rate of fall is 8.15 per cent annually. The Working Group has estimated the potential yield of lobsters to be 5000 tons. This figure was exceeded in 1989 and 1990, after which the production started declining. Being a high value product in the export market its exploitation has not followed any conservation principle, nor is there any attempt to extend the lobsters fishery to deeper regions, particularly off the South-east and South-west Coasts where some rich beds of lobsters are located. It is necessary to undertake further exploratory survey to locate lobster beds and implement stricter conservation measures to save this high value crustacean resource from depletion.

It should be pointed out here that it is the live lobsters which obtain a higher price than the fresh or frozen lobsters. For example, in Japan live lobsters fetch about 4–6 times the price of frozen lobsters, most of which are airlifted from other countries. Unfortunately, the lobsters that are exported from India are in frozen form. The industry should acquire technology for export of this product in live form for better price realization.

Next comes the cooked lobsters and frozen tails which have a good market in the United States and Europe. In the case of frozen tails, though the yield is low (about 30–40 per cent), the price realization is very good because it is considered as a speciality item in these countries.

Crab is the third important variety of crustaceans, but it has not received due attention either in the exploratory survey or for the export market. It is still being considered as a subsidiary fishery despite its having a good export market. Not much of selective gears are used for crabs. It remains a by-catch by shrimp trawls which results in juvenile destruction. As in many countries of the developed world, crabs should be brought under regulatory development and conservation. Attempts should also be made for breeding and farming of crabs, particularly mud crabs, which being larger in size have good export market. There is also a good demand of live crabs in the export market. Crabs could play a role in the diversification of Indian export trade which is more than overdue.

Marine crabs, having reached their highest capture in 1993 (46,667 tons), is nearing the plateau at the present level of efforts (18,081 tons in 1997). While stomatopods and other crustacean products have exhibited consistent growth in their catches, presently they command a share of more than 2 per cent of marine landings in India.

Bombay Duck

This is the second largest contributor to marine fish landings in India. As the name suggests, this fishery is predominantly located in the Bombay–Gujarat coasts. In fact, the West Coast accounts for more than 90 per cent of the total catches of Bombay duck, of which Gujarat contributes about 50 per cent and Maharashtra, including Daman & Diu, contributes the remaining 40 per cent. The other states of the West Coast do not have much share of the catches.

In the East Coast, which shares about 10 per cent of Bombay duck landings, West Bengal's share is about 7 per cent, while the remaining 3 per cent is shared by Andhra Pradesh, Orissa, and Pondicherry. One of the peculiar features of this fish is that it is captured throughout the year, though the landings are highest during the last quarter.

Mechanization and motorization of boats have contributed to a very fast growth in the catches of this fish species, from about 36,000 tons in 1955 to an average of more than 2 lakh tons during the recent years. Presently, it has a share of about 8.50 per cent of total marine landings in India, though on an average, it was maintaining a share of around 6 per cent during the earlier years.

It was surmised during the mid-70s that the annual catch of Bombay duck would stabilize at around 80,000 tons, but we find that actual landings during the recent years have belied the estimate.[4] Total catches have gone up from 103,202 tons in 1987 to 214,217 tons in 1997, with a mean of 149,000 tons, against the official estimate of 104,000 tons yield potential.

Croakers

The next important species of demersal fishery are sciaenids, of which croakers occupy a significant place because of their comparatively large size. Nearly, 90 per cent of croakers land in the West Coast, of which Gujarat alone accounts for nearly 80 per cent. Maharashtra comes next, though with a much smaller share of 8 per cent of the total catches. Karnataka and Goa together share the remaining 2 per cent.

The East Coast has a share of 10 per cent. Orissa occupies first place with a share of about 5 per cent of total catches. Tamil Nadu shares about 3.5 per cent and the remaining 2.5 per cent is distributed among Andhra Pradesh, Pondicherry, and Andaman and Nicobar Islands.

Croakers are predominantly a coastal fishery. Inshore catches are larger in size than the offshore catches which are primarily juvenile. In the early

---

[4] Prasad Raghu R. and P.R.S. Tampi, ibid.

1970s it was thought that this particular species was being overfished and might be facing depletion. While, the recent data do not overtly suggest this, the juvenile exploitation reflected by the fall in the average size of individual fish, as reported in the present study, may be taken as a pointer to overfishing in the near future.

Croaker landings during 1987–97 indicate that its catch may be stabilizing at around its mean catch of 250,000 tons. The variation in annual catches is also low, at around 15 per cent. The present catch quantity of croakers (287,525 tons in 1997) is much above the yield potential of sciaenids of 142,000 tons as estimated by the Working Group.

Pomfrets

This group is another high value demersal fishery of India. These fishes are easy to clean, processed and preserved, and are regarded as a much cherished table fish both domestically and all around the fish-eating world. There are primarily two major varieties of pomfret, namely silver or white and black, of which the former is more liked and fetches a price nearly 2.5 times that of the latter.

The capture of pomfrets is equal between the West Coast and East Coast. In the West Coast, Gujarat has the largest share of about 25 per cent of total catches, followed by Maharashtra (15 per cent). The remaining 10 per cent are almost equally shared by Karnataka and Daman & Diu.

In the East Coast, all the states have some share of the pomfrets. West Bengal tops the list with more than 20 per cent of the total catches, followed by Orissa (13 per cent), Andhra Pradesh (8.50 per cent), and Tamil Nadu (7 per cent).

Nearly 80 per cent of the total pomfret catches are of the silver and white variety. As said before, silver and Chinese pomfrets together comprise the major part of total pomfret landings in India. In fact, we find that the catch in black pomfret has fallen at the compounded rate of about 2 per cent per annum during 1987–97, while silver and Chinese pomfret together have grown by about 7 per cent per annum during the same period. The overall rate of growth for pomfret group as a whole is 2.84 per cent per annum during 1987–97.

Chinese pomfrets are very irregular in submitting to the fishermen's net. The catches vary from zero in a year to nearly 8000 tons in the following year or so. It is also found that when there is a bumper catch of Chinese pomfret in a particular year there is a comparative fall in the catch of silver pomfret. There exists some amount of inverse relationship between the two types of pomfrets.

The Working Group's estimate of the pomfret fishery potential is 54,000 tons including 12,000 tons beyond the depth of 50 metres. The catch data during the past five years suggest that actual landings have already crossed the estimated potential. The mean catch of total pomfrets is 55,344 tons, during 1987–97, which is slightly over the estimated potential. The fluctuation in catches is also low (CV equal to 13.53 per cent).

Ribbon fishes

This species are growing in importance among demersal fisheries in India. These fishes have numerous fine bones which makes them not a highly sought after fish, unlike hilsa shad. But their thin body structure makes them amenable to easy processing through salting, sun-drying etc. They are also easy for storing in bulk quantity. This fish, being low in price, is popular among the poorer sections of the people. This also provides engagement of fishermen during the summer months when other fishes are scarce but ribbon fishes are available in plenty. Although these fishes are mostly captured along the coastal areas by traditional boats, offshore captures by mechanized boats are found to be of longer varieties, even up to one metre length, as against 300–500 mm length captured along the coastal areas.

Ribbon fishes are mostly caught along the East Coast. Tamil Nadu accounts for the bulk of the capture, followed by Andhra Pradesh. The West Coast's contribution is not very significant.

Catches of ribbon fish are on the increase (from 61,026 tons in 1987 to 160,158 tons in 1997), due primarily to offshore exploitation beyond 50 metres depth by mechanized vessels, where the yield potential is high—about 216,000 tons—as compared to a potential of 95,000 tons within 50 metres depth, which is mainly exploited by traditional boats.

The share of ribbon fishes in the total marine landings has gone up from 3.17 per cent in 1987 to 6.37 per cent in 1997. This is due to the rising demand of ribbon fish in the export market.

Cephalopoda

Squids, cuttlefish, and other molluscs that constitute the cephalopoda group are traditionally the next important seafood export of India, besides prawn and shrimp. But India's share of production and export of cephalopoda as compared to other Asian countries is very low. In fact, Asia shares more than 60 per cent of the world cephalopoda production. Japan tops the list with more than 25 percent share of the world production, followed by the Republic of Korea (17 per cent). India's share is only about 3 per cent.

India's cuttlefish exports are mainly to Japan, China, Hong Kong, Thailand, and Italy. Hong Kong and Thailand import cuttlefish from India mainly for re-exporting. The export to Italy is nominal. Indian squids products have good demand in Australia, France, and other European countries, besides Hong Kong and China. India does not have much of a domestic demand in cephalopoda, which has traditionally made this species export-oriented. Cephalopoda consumption is showing a rising trend, particularly in Asian countries. With due attention, India could emerge as a leading exporter of this seafood.

The West Coast accounts for nearly 88 per cent of the squids and cuttlefish production. Kerala tops the list with about 40 per cent of total catches of squids and cuttlefishes, followed by Gujarat (28 per cent) and Maharashtra (12 per cent). Karnataka accounts for about 8 per cent of the total catches of these two cephalopoda.

In the East Coast, Andhra Pradesh, Orissa, and Tamil Nadu together share about 10 per cent of the total catches of squids and cuttlefishes. Pondicherry has a small share.

The production of other molluscs vary wildly from coast to coast and state to state. They occur periodically in Andhra Pradesh, Pondicherry, Orissa, and Tamil Nadu in the East Coast, and Karnataka in the West Coast.

Although cephalopoda catches in India have grown at a cumulative rate of around 13 per cent per annum during 1987–97, the actual spurt in the growth has occurred since 1991 when the catches nearly doubled that of the earlier year. The present trend is towards a stabilization of catches at around 100,000 tons, though molluscs catches as a component of cephalopoda continue to vary wildly. The Working Group's estimate of the cephalopoda potential is 71,000 tons, which has already been exceeded in 1992 itself (81,573 tons).

Among the cephalopoda, squids are receiving greater attention in the speciality segment of the export market. Squids have a higher protein content than other commercially important fishes and, though these fishes are not so much consumed in India, in Japan these are considered as luxury foods.

The world demand of squids is on the rise. The oceanic catches may not be able to satisfy this rising demand. In the advanced world, attempts are being made to produce cultured squids. India should also move towards squid culture to participate more effectively in the growing export market of squid products.

Seer fishes

These fishes closely resemble the king mackerels of the American coast. Being rich in protein they have good demand in the United States and Europe. These are exported mainly in whole IQF form (Individual Quick

Freezing) with or without gutting. Indian seer fishes have good demand in the markets of Singapore, Japan, and Kuwait. The international quality standards for this fish product are very stringent.

Seer fishes are widely distributed along both the West Coast and the East Coast. Almost all the coastal states have some share in their catches. However, the West Coastal states have the larger share, about 57 per cent of total catches. Gujarat comes first, accounting for about one-fifth of the total, followed by Kerala, Maharashtra, and Karnataka.

Along the Eastern Coast, Tamil Nadu leads with another one-fifth of total catches, followed by Andhra Pradesh and West Bengal.

Except for the bumper catches of 1995 and 1996, the capture of seer fish in India is moving to around 39,700 tons on an average, which is close to the Working Group's yield estimate of 42,000 tons.

Perches

This fish species comprise a large number of various genre of fishes widely distributed along all the coasts. In India, perches comprise about fourteen varieties of which rock cods, snappers, pig face breams, and thread fin breams are the most important. The volume of their catches vary widely from one year to another. Some of the perches, like rock cods and snappers, are found in abundance in the rocky regions and coral reefs and other seagrass beds of coastal waters but they are difficult to catch, except with special devices. Hence, their potentials are not fully exploited.

Although the West Coast accounts for the largest catches of perches, it is difficult to establish any percentage distribution either between the two coasts or among the coastal states because of a wide variation in their occurrences. A state may have a bumper catch in one year followed by almost zero catch in the next year. For example, in 1996 the catches were 133,565 tons, which came down to 80,915 tons in 1997.

Perches, considered as a minor fishery in the 1970s, have now grown much faster in volume with a cumulative growth rate of 8 per cent per annum. Presently, perches claim a share of more than 3 per cent of the total marine landings.

The Working Group's estimate of the potential yield of perches (predominantly threadfin breams) is 239,000 tons, of which 125,000 tons are coming from beyond 50 metres depth. It appears that a large part of this fishery has remained unexploited, particularly beyond 50 metres depth.

Elasmobranchs

This fishery in India is principally composed of sharks, rays, and kates. India is among the first five leading producers of elasmobranchs in the world. The value of these fishes is comparatively low, because of their peculiar smell. However, these fishes are utilized both in cured and fresh form. The other profitable usage is reduction for oil. Shark-liver oil has already established itself as a substitute of imported cod-liver oil. A good market for Indian shark fins exist in the Far-Eastern countries. These fishes have also good usages in fish meals and manures.

Among the elasmobranchs, the whale shark is emerging as a distinct fishery, particularly in Veraval of Gujarat. The individual size of this fish vary between 4.5 and 12 metres with weights between 2 and 8 tons. Till recently, only the fins and livers of the fish were utilized and the carcasses used to be discarded. However, a good market for whale shark meat has emerged in Korea, Taiwan, Sri Lanka, and Singapore. Presently, India's export of whale shark meat is about 200 tons, but there is much scope for development of this market. The meat is exported primarily in frozen form, but some part is also sold in salted and dried form. The other usage of this fish is reduction for oil, though because of the crude methods employed for extraction of oil, the price it fetches is generally small.

Shark cartilages[5] are also reported to have properties that curtail the growth of blood vessels and thus restrict the spreading of diseases like blood cancer. It is expected that many other usages of different parts of the shark will be discovered in the future and the sharks will further grow in importance. There is now much scope for this hitherto neglected fishery.

While smaller elasmobranch fishes are captured near the coast, large sizes are generally found in distant waters. There is scope for increasing catches in offshore waters with the help of mechanized boats.

The East Coast accounts for the largest share of elasmobranch catches (more than 60 per cent). All the states of the East Coast have some share in the catches, with Tamil Nadu leading with a share of about 20 per cent of total catches, followed by West Bengal (about 16 per cent). Andhra Pradesh and Orissa follow immediately thereafter.

Along the West Coast, Gujarat has the largest share (more than 15 per cent), followed by Maharashtra (about 8 per cent). All the other states have some share in the total catches.

A study of the catch data of elasmobranchs from 1987 to 1997 indicates that the production of shark is not increasing at all, rather the trend is on the decline. Even at the aggregate level, the production of elasmobranchs is showing a deceleration in the rate of growth. Even if we exclude the year 1997, when it reached the lowest level at 57,996 tons, we find that the compounded rate of growth is very small, at 0.65 per cent per annum.

[5] *Indian Express*, 'Shark Cartilage', 21 August 1996.

The Working Group has estimated the potential yield of elasmobranchs at 168,000 tons, of which 103,000 tons come from a depth beyond 50 metres. The catch data during the past ten years indicate that even half of the potential is not being exploited.

Catfishes and Silver Bellies

These two groups of fish species have gained importance in the present decade. Catfishes are carnivorous in nature and feed on small fishes. These are distributed along both the coasts of India and are available both within and beyond 50 metres depth. Although economically important by volume, this fish-group generally commands a low price in the market. In India there are about six species under this group, known popularly as Asian catfishes. The most important among them are the white catfish and black catfish, weighing between 2 kgs and 9 kgs. There are also some small varieties weighing between 500 gms and 1.25 kgs. This group of fishes is available in all the continents of the world, except Oceania and, is known by the continent's name. These are also popular fish products consumed throughout the world. The growing demand has led to aquaculture development in these fishes in America and Europe. But the true commercial importance of these fishes is yet to be fully appreciated by the Indian seafood industry.

Catfishes appear to have reached a stability in production at around 75,000 tons after the high growth of 90,000 tons during 1991–4. The cumulative rate of growth for the whole period is 2.02 per cent per annum with a low level of fluctuations. The Working Group's estimate of the potential for this fishery is 123,000 tons, with 63,000 tons coming from beyond 50 metres depth. It is likely that the potential beyond 50 metres depth is not being exploited.

Silver bellies are small-sized fishes, found in large shoals near the coast, that are mostly captured by traditional crafts and gears and consumed locally. These are mostly found in the East Coast which account for nearly 80 per cent of total catches. Tamil Nadu alone has a share of more than 60 per cent. Andhra Pradesh and Orissa come thereafter. In the West Coast, Karnataka and Kerala together account for about 5 per cent of the catches. Other states have some small shares. Besides their usage in human consumption, silver bellies are also used for fish meal production.

Silver bellies have experienced a severe fall in catches in 1997. If this year is excluded, the compounded rate of growth comes to 7.09 per cent per annum, which is quite good considering the comparatively low level of fluctuations in catches (CV equal to 26.47 per cent with a mean of 50,558 tons). The estimated potential of silver bellies is 86,000 tons with only a small quantity of 4000 tons coming from beyond 50 meters depth. It appears that there is still scope for exploitation of full potential of these fishes.

Other Minor Fisheries

Besides the major fisheries discussed above, there are several other minor fisheries which are in different stages of development in India. Landings from minor fisheries together constitute about 5 per cent of the total marine catches, whereas the unspecified or miscellaneous group presently has a share of about 10 per cent.

Of the minor fisheries, flatfishes, comprising halibut, flounders, and soles, have gained considerable importance during the past fifteen years. On an average, this group of fish species now claims nearly 2 per cent of total marine landings. The growth in catches is primarily due to soles which constitute more than 90 per cent of the landing of flatfishes. Though halibut and flounders have a good export market, their production along the Indian coasts is very small. Soles have a worldwide market. As the production of sole is declining in the developed countries but their demand is on the rise, India has much scope in the export market of this product.

PRODUCTION EFFICIENCY INDEX

The commercial importance of a fish species or a group of species from the point of view of the seafood industry, particularly those engaged in the export of fish products, depends upon the volume of catches, rate of growth, and stability in catches.

Unless the volume of catches of a fish group is sufficient, commercial investment in that fishery may not yield the desirable return to sustain it on a continuous basis. A fishery may be commercially very important but if the volume of landings is low, it may be unwise to make substantial investment in that fishery. This important point is often missed. The commercial importance of the fishery, particularly in the export market, sidelines the basic economic tenets. As a result, a misallocation of resources takes place, resulting in the creation of excess capacity. In our model, volume is measured by the percentage share of a particular fish group in the total marine landings in a given cut-off year, which has been taken here to be 1996.

The next important variable is the rate of growth in catches, as investment planning is always done on the basis of future growth. If the volume and its growth rate are very high the scale of investment will also be high, because in such a situation, the industry may like to cash in on the future prospect of the fish product. There may, however, be cases when a high rate of growth may be due to a low catch-base of a particular fish group in the first year of the period under consideration. However, this is moderated by the volume of catches as measured by the share of total catches discussed above. For our model, the rate of growth is taken as the cumulative growth rate for the period 1987–96.

A particular fish group may have a good volume and rate of growth but it may still lack commercial viability if the year to year fluctuation in catches is very high. Uncertainty in raw materials (fish) supply makes the investment highly risky both from the perspective of production planning and sustainable sales planning, particularly in the export market. For this reason, a manufacturer (processor) would always like to ensure a stable source of raw materials. In other words, he prefers those fish species which have a stable growth rate in catches (this, however, does not mean stagnancy).

The fluctuation level of catches is measured, as before, by the coefficient of variation (CV) for the given period. However, a high CV may often indicate the liveliness of a particular fish species. It may be due to migratory behaviour of the species or the inherent biological nature, but commercial exploitation of the species becomes risky from a manufacturer's point of view. The stability consideration is factored in the model as (1–CV). The period chosen is 1987–96.

For the purpose of explaining the model we have ignored the year 1997 because it was an unusual year in many respects, as explained in Chapter 10.

The model is finally presented below as a multiplicative function of volume (V), rate of growth (g), and stability (1–CV) of catches of a particular fish group:

Product Efficiency Indicator (PEI) = V x g x (1–CV)

Product Efficiency indicators of all the major fish groups, thus calculated, are then used for developing the Production Efficiency Index. This is done by dividing the Product Efficiency Indicator (PEI) of an individual fish group by the summation of all such PEIs and then converting them into percentages. Individual fish groups are then ranked in the descending order of the Indexes. This is done in the Table 8.3.

We have seen earlier that Indian mackerels and carangids group as a whole are subject to heavy fluctuation in catches which would otherwise have made them less suitable for commercial exploitation, but their high rates of growth coupled with the considerable volume of catches have moderated the fluctuation parameter and made them rank first and second in terms of production efficiency. On the other hand, oil sardines, presently have a good volume but the other two parameters, namely growth rate and

TABLE 8.3
Production Efficiency Indexes of different fishery groups

Rank	Fish species group	Cumulative growth rate (1987–96)	Share of total marine landing (1996)	Levels of fluctuations (1 − CV)	Product Efficiency Indicator (col. 3 x 4 x 5)	Production Efficiency Index
(1)	(2)	(3)	(4)	(5)	(6)	(7)
1	Carangids	0.2171	0.0646	0.4286	0.00601	14.00
2	Indian mackerels	0.1196	0.0665	0.6122	0.00487	11.34
3	Ribbon fishes	0.1060	0.0582	0.6793	0.00419	9.76
4	Penaeid prawns	0.0733	0.0698	0.7991	0.00409	9.52
5	Perches	0.1441	0.0466	0.5805	0.00390	9.08
6	Croakers	0.0446	0.0944	0.8484	0.00357	8.31
7	Bombay duck	0.0603	0.0646	0.8498	0.00331	7.72
8	Other clupeoids	0.1484	0.0453	0.4715	0.00317	7.38
9	Cephalopoda	0.1470	0.0324	0.5145	0.00245	5.70
10	Silver bellies	0.0709	0.0268	0.8006	0.00152	3.54
11	Tunas*	0.0662	0.0185	0.8334	0.00102	2.37
12	Flatfishes	0.1160	0.0143	0.5972	0.00099	2.30
13	Other crustaceans	0.1039	0.0211	0.4479	0.00098	2.28
14	Seer fishes	0.0526	0.0216	0.7831	0.00089	2.07

(Contd.)

Table 8.3 (contd.)

Rank	Fish species group	Cumulative growth rate (1987–96)	Share of total marine landing (1996)	Levels of fluctuation (1 – CV)	Product Efficiency Indicator (col. 3 x 4 x 5)	Production Efficiency Index
(1)	(2)	(3)	(4)	(5)	(6)	(7)
15	Pomfrets	0.0456	0.0223	0.8580	0.00087	2.03
16	Non-penaeid prawns	0.0132	0.0401	0.9055	0.00048	1.12
17	Other sardines	0.0179	0.0279	0.8143	0.00041	0.95
18	Cat fishes	0.0175	0.0249	0.8607	0.00037	0.86
19	Mullets	0.0751	0.0072	0.6685	0.00036	0.84
20	Lizard fishes	0.0932	0.0056	0.5958	0.00031	0.72
21	Elasmobranchs	0.0065	0.0248	0.8680	0.00014	0.32
22	Eels	0.0403	0.0040	0.8264	0.00013	0.30
23	Oil sardines**	(–) 0.0240	0.0508	0.5739	(–) 0.00070	(–) 1.63
24	Anchovies	(–) 0.0439	0.0130	0.6634	(–) 0.00038	(–) 0.88
	TOTAL				0.04295	100.00

*   For the period 1987–96
**   For the period 1988–97

Source: *Handbook on Fisheries Statistics*, Ministry of Agriculture, Fisheries Division, Govt. of India, New Delhi, 1996 and *Statistics on Marine Products Exports*, The Marine Products Export Development Authority, Cochin, various issues.

fluctuation in catches, are so adverse that this fish species ranks second lowest in the list.

Under the prawn group, penaeid varieties rank fourth, as all the efficiency parameters are good, but the non-penaeid varieties rank as low as sixteenth because the first two parameters are very poor, though there is stability in the catches. In fact, this stability indicates a stagnation in the catches of non-penaeid prawns.

Production Efficiency Indexes (PEI) can be used to derive strategic focus by the seafood industry engaged in commercial exploitation of different fish species. The indexes will be useful for developing a portfolio of fishes for investment and market planning. At the national level these can be used for allocation of developmental resources among different fisheries.

PROJECTIONS FOR THE FUTURE

While discussing individual groups of fish species we have seen that many of them suffer from heavy fluctuations in catches, which made it difficult for us to develop statistically significant regression equations. But at the aggregate level, and over a longer period of time the fluctuations are smoothened out which makes it amenable to developing appropriate predicting equations. Table 8.4 presents data for both marine and inland fish landings from 1980–97 primarily for this purpose.

Figure 8.1 gives a graphical presentation of the data in Table 8.4.

It can be observed from Table 8.4 that the growth in the catches from the marine sector is slower than in the inland sector. The cumulative growth rate of the marine sector is 2.70 per cent per annum, as compared to the growth rate of 5.64 per cent per annum achieved by the inland sector. The Working Group's estimate of potential yield from the marine sector is 3900 thousand tons. In terms of this estimate, India has just about exploited 65 per cent of her marine potential. However, if we look at the catches of marine sector from 1992, we may notice that there is a stagnancy at around 2600 thousand tons. The stagnancy is primarily due to the concentration of Indian fishing in coastal waters and the reluctance of even mechanized vessels to go beyond 50–75 metres depth. We have seen before that this has resulted in full exploitation and, in some cases, overfishing, particularly in the West Coast. A large potential of fishery resources beyond 50 metres depth remain unexploited. It may not be too wrong to say that the declaration of the 200-mile Exclusive Economic Zone (EEZ) has not benefited India much, but has deprived the world of potential fishery resources.

The stagnancy in catches has given rise to a lower coefficient of variation in marine catches, as compared to inland water fish production. At the

TABLE 8.4

Marine and inland water fish landings in India

(in thousand tons)

Year	Marine	Inland Water	Total
1980	1555	887	2442
1981	1445	999	2444
1982	1427	940	2367
1983	1519	987	2506
1984	1698	1103	2801
1985	1716	1160	2876
1986	1713	1229	2942
1987	1647	1301	2948
1988	1853	1335	3188
1989	2158	1402	3560
1990	2202	1536	3738
1991	2387	1710	4097
1992	2604	1789	4393
1993	2556	1995	4551
1994	2673	2097	4770
1995	2604	2242	4846
1996	2870	2283	5153
1997	2513	2380	4893
Mean	2063.33	1470.29	3507.18
SD	478.50	454.11	925.85
CV	23.20	30.88	26.40

aggregate level of catches of both marine and inland water sector, we find that the fluctuation is about 25 per cent around a mean catch of 3500 thousand tons. During the recent period (1989–97) the CV has come down further to 11.58 per cent.

We have made regression analysis of the catch data of both marine and inland water sectors and also of the total catches separately. Regression parameters of the equations are given in Table 8.5. For the inland water sector, the regression equation is developed on the data period 1980–96.

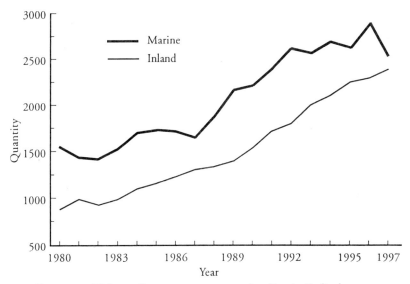

Fig. 8.1: Fish, molluscs, crustaceans etc. landing in India by sector
(in thousand metric tons)

TABLE 8.5
Regression parameters of catches from marine and inland sectors

Sector	Constant(a)	(b)Coefficient	SE(b)	t. value	R²
Marine	1230.65	87.65	7.17	12.22	0.90
Inland Water	655.00	90.59	5.07	17.86	0.96
Total Catches	1844.38	184.75	10.27	17.99	0.96

Note: At 1% level of significance.

Regression equations given in Table 8.5 reveal that catches from both marine and inland water sectors are increasing at a decreasing rate, but the rate of deceleration is higher in the inland sector.

On the basis of above regression parameters we have made a projection of fish production in India for the period 2000–5 in Table 8.6.

TABLE 8.6

Projected sector-wise and total fish production in India

(in thousand tons)

Year	Marine	Inland Water	Total
(1)	*(2)*	*(3)*	*(4)*
2000	3071	2557	5724
2001	3159	2648	5909
2002	3247	2739	6094
2003	3334	2829	6279
2004	3422	2920	6463
2005	3510	3010	6648

Note: Projected total catches in col. 4 will not tally with the aggregate of col. 2 and 3 because total catches are projected separately.

DEMAND–SUPPLY GAP

We have indicated earlier that due to a rise in income of the Indian population, domestic consumption of fish is on the rise. The fish consumption data published by the Food and Agricultural Organisation reveal that per capita consumption of fish in India has gone up from 3.3 kg in 1988 to 4.4 kg in 1995. That is, the cumulative growth rate in per capita consumption is 4.1953 per cent per annum. We also find from the same data source that population in India has grown cumulatively by 1.9468 per cent per annum during 1990–5. Assuming that both the two trends continue in the future, domestic consumption of fish will rise to the levels projected in Table 8.7, which also gives the projected demand–supply gap of fish products in India.

Table 8.7, along with Table 8.6, indicates that given the existing trends of fish production in the marine and inland water sectors, India will face a supply shortage from the year 2003 even to meet domestic consumption demand, not to speak of availability of surplus for export. We may recall that in this respect India's position is similar to that of China.

Against this, it may be argued that exportable fish species from India are not generally those which are domestically consumed or are available at affordable prices (shrimps) for domestic consumption. Table 8.8 lists the major fish species exported presently (1997–8) from India, ranked in terms of their percentage of production.

TABLE 8.7
*Projected domestic consumption and demand-supply gap
in fish products*

(in thousand tons)

Year	Projected domestic fish consumption	Projected fish production (col.4 of table 147)	Demand-supply gap (col.3 – col.2)
(1)	(2)	(3)	(4)
2001	5745	5909	164
2002	6065	6094	29
2003	6500	6279	(221)
2004	6955	6463	(492)
2005	7315	6648	(667)

Note: Figures in brackets denote supply gap.

TABLE 8.8
*Rank of major fish species exported from India by percentage of
production (by tonnage)*

Rank	Name of fish species	Tonnage export as % of production (approx.)	Share of exports (%) on comparable base
1	Cephalopoda	84	15.04
2	Ribbon fish	76	23.72
3	Reef cod	38	0.40
4	Shrimps	37 (80)	44.16
5	Pomfrets	20	2.30
6	Seer fish	14	1.16
7	Eel	12	0.24
8	Snappers	9	0.20
9	Mackerel	4	0.70
10	Tunas	1	0.15

Table 8.8 indicates that though the level of domestic consumption of cephalopoda and ribbon fish is low, the same cannot be said of other fish species. At the same time, it is also true that these two fishes constitute nearly 40 per cent of the fish exports of India (by tonnage). In case of shrimps, the percentage of exports to production figure given in Table 8.8 does not give a true picture because the weights are of different bases. While production figures represent weights in whole form, the export figures represent weights in whole form as well as in beheaded and peeled forms. The difference in weight between the two forms is quite substantial. On a weighted average, one ton of beheaded/peeled form can be taken as equivalent to 2.2 tons of shrimps in whole form [conversion factors for weights are taken as: whole form (1); beheaded only (2/3); beheaded and peeled form (1/3)].

Going by the above calculation, comparable tonnage exports of shrimps are about 225 thousand tons and total exports, about 510 thousand tons for the year 1997–8. (We have ignored here loss of weights due to cleaning etc. of other exportable fish species.) Hence, the actual percentage of shrimp exports on production comes to around 80 per cent. That is, at present about 20 per cent of shrimp production (which are generally not found to be exportable due to their sizes, nature of the species, or quality standard) are consumed domestically. If we add up the recasted percentage share of export of cephalopoda, ribbon fishes, and shrimps, the aggregate percentage comes to around 80 per cent. The remaining 20 per cent is coming from those fish species whose domestic demand is very high. In other words, if the projected rise in domestic consumption of fish is materialized on the face of a deceleration in the rate of growth of fish production, then at the minimum the fish segments representing this 20 per cent of exports will be affected, which in turn will affect the diversification efforts of the Indian seafood export business. Besides, as we have mentioned earlier, the domestic demand of shrimps may not remain stationary at 20 per cent of production. It is expected to rise owing to the increasing level of purchasing power of the upwardly mobile middle class of India who will demand export quality shrimps for home consumption.

As India cannot afford to lose the precious foreign exchange that she earns from seafood export, and as the fish export trade is fairly well established in India's economy, there will invariably be a drawdown on the existing fish supply, at least at the existing rate of total fish production. This will further reduce the availability of fish for the domestic consumption. As a result, the domestic price of the relevant fish species will sky-rocket, which may ultimately lead to public resistance.

What is necessary, therefore, is to break the existing trends in production both in the marine sector, by emphasizing on the development of a mechanized and diversified fleet for offshore fishing and faster development of coastal water culture, and in the inland water sector by adopting multiculture practices.

Our investigation has revealed that in India, aquaculture has become synonymous to shrimp culture. This monoculture orientation has resulted into biological disequilibrium in our water bodies which has led to many disasters and environmental problems, leading ultimately to a ban on aquaculture ordered by the Supreme Court. A reorientation towards multiculture is, therefore, the order of the day. A study should be made to develop a structure of compatible and complementary fish species that can be grown at different layers of the water. This approach will also diversify the risk and protect the aquaculturist against sudden failure of one species without, at the same time, reducing his aggregate return on investment. The future aquaculture policy should be directed towards this diversification.

ENVIRONMENTAL ISSUES

At the same time we must draw attention to the worldwide environmental problems created by unscientific and uncontrolled expansion of aquaculture activities, as discussed in Chapter 3. India has also suffered from the same phenomenon. As a result, in 1996 the Supreme Court came out with an order for demolition of all those aquaculture firms in the Coastal Regulation Zone (CRZ) that are creating pollution of seawater by their effluent discharge. This was followed up by another order directing the central government to set up an Aquaculture Authority. A Bill was subsequently introduced in the Parliament to formally set up an Aquaculture Authority which will, *inter alia*, have powers to:

(a)   prescribe regulations for the construction and operation of aquaculture farms within the coastal area;

(b)   inspect aquaculture farms with a view to ascertaining their environmental impact;

(c)   grant licences to aquaculture farms and also to order removal or demolition of any aquaculture farm which is causing pollution.

While the setting up of an Aquaculture Authority is most welcome for the regulated development of aquaculture farms in India, it appears that the environmental problems created by aquaculture activities in India have been overblown by the environmentalists. (The National Fishworkers Forum had staged a hunger strike on 9 January 2001 against the introduction of the Bill in Parliament on the ground that the Bill was a bid to bypass the Supreme Court order for demolition of all aquaculture farms in the vicinity of the coast to prevent sea pollution.) As early as 1994, the Marine Products Export Development Authority (MPEDA) commissioned the National Environmental Engineering Research Institute (NEERI) to make a study on the environmental management of aquaculture activities in India. After analysing a number of samples, NEERI reported, *inter alia*, that the quality of shrimp farm effluents, when compared with wastes from other potential sources of pollution, indicates that the pollution potential of shrimp farm effluents is considerably less than that of domestic or industrial wastewater. The actual quality of shrimp farm effluents is also less noxious as compared to other sources of coastal pollution. Despite these findings, it can be said that the regulatory measures proposed in the Aquaculture Authority Bill are timely to prevent future disasters, as experienced by some countries of South East Asia.

DEEP-SEA FISHING

One other way by which India could increase fish production in the marine sector is by putting renewed emphasis on deep-sea fishing beyond 50 metres depth. The imperative also comes from the export sector, as many of the fish species beyond 50–75 metres depth do not have much of domestic consumption preference, and hence these can be directly exported without affecting the domestic supply of fish. The Working Group's estimate of the potential yield beyond 50 metres depth is 16.9 lakh tons of which the following groups of fish species claim the major share (yield in lakh tons): tunnies 2.42; oceanic tunas 2.09; carangids 3.04; ribbon fishes 2.16: perches 1.25; elasmobranchs 1.03.

The above six groups of fishes constitute about 71 per cent of the potential beyond 50 metres depth. While analyzing the catch performance of individual fish groups, we have seen that many of their exploitation levels have not gone beyond 50 metres and, as such, some of them are reaching a stagnation level. Among these six groups, tunas and carangids top the list in terms of commercial value. The Working Group observed that of the total tuna resources of 4.88 lakh tons (including tunnies, both inshore and offshore), only about 7 per cent were being exploited presently, and in the case of carangids the exploitation level was about 25 per cent of the potential. Hence, in order to break the present trend in marine fish production, it is necessary to develop an appropriate strategy for exploiting these resources.

The Working Group, in 1990, aimed at exploiting 50 per cent of the resource potential beyond 50 metres depth by introducing 2630 mechanized vessels classified under three categories, namely 25–30 metres trawler suitable for pelagic and bottom trawling; 12–30 meters gill-netter-cum-longliners; and 12–20 metres, 25–30 metres, and 36–40 metres and purse-seiners of 36–40 metres for exploiting oceanic skipjack tunas. These measures were meant primarily for exploiting the high value resource potential, as mentioned above. For low value fishes, the Working Group recommended the introduction of low cost, fuel efficient vessels.

The Working Group's recommendations were an extension of the deep-sea fishing policies adopted by the government earlier. As early as the 1970s, the government recognized the need for deep-sea fishing and approved the import of 30 deep-sea shrimp trawlers from Mexico to boost up fish production from the deep sea. But the impetus provided by the government was slow to catch up, and hence in 1987 the government declared a new deep-sea fishing policy which, among other things, allowed the charter of fishing vessels and the liberalized import of new and second-hand vessels. The policy also laid down permissible sizes and types of deep-sea fishing vessels and encouraged joint ventures with foreign collaborators for the exploitation of deep-sea resources.

What was missed, however, was the bio-economic imperatives of any broad scale fishing venture. Deep-sea fishing (DSF) policies were totally shrimp oriented, and worked well for both the entrepreneurs and the biological system of shrimp resources from 1970 to the early part of 1980s. The fleet strength was growing slowly, and by 1983 the fleet size was only 68 trawlers at Visakhapatnam (Vizag), the centre for deep-sea fisheries in India. But the government not being satisfied with slow pace of DSF, due perhaps to the country's need for increasing exports, embarked on faster development of DSF, first by augmenting and adding incentives for the acquisition of deep-sea vessels, which ultimately culminated in a policy declaration in 1987, as mentioned above. In response to these policy initiatives, the DSF fleet grew at a very fast rate, both by number and fishing power. By 1991, the fleet strength had grown to 180 vessels with a total fishing power of more than 84,000 horse power. The germ of sickness, which was evident in 1987 itself, got a boost by the new deep-sea fishing policy in the same year, and by 1991 it had became a killer. The average catch per trawler, which was about 25.5 tons between 1981 and 1986, came down to an abysmally low figure of 7 tons. Even allowing that DSF operations were affected by crew strikes during 1989–90, the average catch per trawler could never have been more than 8 tons. At this level of catches, the DSF operation simply became unsustainable, resulting in the erosion of net worth and consequent default on servicing of loans taken from financial institutions and banks. Virtually, the entire trawling industry became sick.

Murari Committee

The Technical Committee on deep-sea fishing (DSF) appointed by the Government of India under the chairmanship of P. Murari recommended, among other things, a substantial financial restructuring of the DSF industry.

Although the committee recommended liberal concession to defaulting DSF units, particularly under the one-time settlement option, and several firms have gone in for such settlement of their debt by collecting funds from several sources, including private loans, real rehabilitation of these units is not taking place because, having put in funds for one-time settlement, many of them are suffering from a shortage of working capital to undertake a fishing voyage. It is also true, at the same time, that all such units or their owner(s) have not gone bankrupt, as is revealed in private talks with a cross-section of entrepreneurs who have exercised their debt-relief options and are carrying on their operations, though on a limited scale. The discussions also revealed that a good number of such entrepreneurs have valuable house properties which they are prepared to offer to banks as collaterals for working capital loans. However, banks are found to be not prepared to grant working capital loans to them because they have lost confidence in the ability of these units to make a turnaround because of their 'proven failure'.

The Murari Committee did mention that 'working capital requirement for the industry should be provided through a revolving fund with commercial banks, for which refinance could be arranged by an appropriate agency', but no methodology was proposed to allay the anxiety of bankers who simply 'do not want to create further NPA [non-performing assets] for their banks'.

The anxiety of the bankers is justified because by now it is known that shrimp trawling has lost its viability. There is simply not enough shrimp in the Indian seas due to competitive pressure of both the over capacitated DSF fleet and small-scale motorized fleet, who having seen the bumper catches by the DSF vessels in the later part of the 1970s moved in to have a share of the cake. The resolution, therefore, lies in getting away from the shrimp orientation of DSF venture and moving towards diversification.

Diversification of Deep-sea Fishing

That the diversification was the need of the hour was advocated forcefully by Giudicelli as early as 1992 when he argued that the DSF policy should, as a first step, be directed towards decreasing the pressure of deep-sea vessels on the penaeid shrimp stock through retargeting a substantial portion of its catching power on other resources.[6] But at the same time, he lamented that in many countries, where industrial (DSF) fisheries started with penaeid shrimp trading in shallow waters, the trawler-men used to become complacent and, being accustomed to the easygoing style of shallow water fishing, they do not seem to be excessively keen on oceanic longlining operations, requiring intensive and fast deck work at great distances from the shores. Unfortunately, the Indian DSF establishment also yielded to this syndrome, so much so that even this warning was not heeded to by the sector. Only when the fire has broken out have we seriously begun talking about diversification of DSF operations. The Working Group's estimate of the potential yield of fish resources beyond 50 metres' depth is 16.41 lakh tons, of which about 16.70 per cent is from beyond 200 metres depth. About 64 per cent of this potential is located in western waters; 30 per cent in the eastern waters and the remaining 4 per cent in Lakshadweep and Andaman & Nicobar waters.

According to Giudicelli's estimate, out of the above potential, 11.25 lakh tons might be available for commercial exploitation, of which 4 per cent belong to very high value or high value category, 10 per cent under medium value category, and the remaining 86 per cent come between low and very low category of commercial value. He opined that even if only 14 per cent of this potential (comprising very high to medium category of commercial value) could be exploited it would increase the marine fish production by 7 per cent, but that would be enough to increase the export value of this sector by more than double the present value.[7] This is so because the fish species coming under these categories do not have much of a domestic market but they attract high value in the export market. One such fish species is tuna, to which we now turn.

TUNA FISHING

Although belated, a Round Table Conference on Tuna Fishing was convened by the Association of Indian Fishery Industries at Visakhapatnam on 14 June 1999 to explore the possibility of diversifying a section of the deep-sea trawlers to tuna fishing. The Conference, *inter alia*, proposed the conversion of about 30 per cent of the existing deep-sea fish trawlers for undertaking monofilament tuna long lining. Dixitulu, whose cost-benefit models for such diversification were adopted by the Conference, has favoured

[6] Giudicelli, M., *Study on Deep Sea Fisheries Development in India*, Food and Agriculture Organisation, Rome, 1992.
[7] Giudicelli, M., Ibid.

the installation of one or more additional systems on existing shrimp trawlers, instead of acquiring new vessels for this purpose.[8]

Oceanic tuna resources, as estimated by the Working Group, is 242,000 tons, of which Andaman & Nicobar Islands have 100,000 tons and Lakshadweep Islands have 50,000 tons. We have seen earlier that the present production from the last two regions is about 15,000 tons, leaving unexploited resources to the tune of 135,000 tons from these two regions alone. The total tunas landing in India as on 1997 are about 46,500 tons. Assuming that 20,000 tons are captured in the coastal areas, then the available oceanic tuna resources would be around 225,000 tons (net of tonnage presently captured in Andaman & Nicobar and Lakshadweep Islands). Even if half of this available potential could be captured, India's export of fish products could be doubled in value terms.[9]

Before we evaluate the financial viability of tuna diversification, we should analyse the findings of our survey among trawler owners. The findings of the survey are based on a structured questionnaire, formal discussion, and informal private talks.

FINDINGS OF SURVEY

Trawler operations are predominantly corporatized. About 70 per cent are public limited companies and the remaining 30 per cent are private limited companies. It is found that 54.55 of the companies own more than one trawler; such ownership varies between two and four trawlers.

On the basis of the total fish capacity reported by the trawler companies, the average net fish capacity of a trawler is found to be 25.68 tons.

The fishing season is reported uniformly as June–March. Nearly 82 per cent of the respondents reported that they go for fishing for 180 days which they complete by six trips. That is, on an average, one trip consists of 30 days. The annual net fish capacity of a trawler could, therefore, be taken as 25.68 x 6 = 154.08 or, say, 154 tons.

Nearly 92 per cent of the trawlers remain on shore for 90 days, 8 per cent for 150 days, and another 8 per cent for 224 days. The reasons ascribed for being idle are given in Table 8.9.

It is observed that besides repairs of boats and nets and bad weather, which are the common features for any ocean going fishing boat, the next important reason for the boat remaining idle is lack of working capital,

[8] Dixitulu, J.V.H., 'Addition of Monofilament Tuna Longlining System on Shrimp Trawlers and its Indicative Economics', paper presented in the Conference cited above, Visakhapatnam, June 1999.

[9] Calculated at Rs 5.16 lakh per ton (as estimated by Dixitulu) on 112,500 tons = Rs 5805 crore.

which must be a new phenomenon being experienced by the trawler operators during the past five years owing to erosion of net worth and banks' unwillingness to provide any further working capital.

TABLE 8.9
Percentage distribution of reasons ascribed for trawlers remaining idle

Reason ascribed	Per cent of total responses
Repairs of boats and nets	30.00
Festivals	5.00
Bad weather	27.50
Sickness of the crew	10.00
Conservation period	5.00
Lack of working capital	22.50
Total	100.00

The average size of the crew per trawler boat is reported to vary between 12 and 14. About 64 per cent of the respondents pay between Rs 8 lakh and Rs 9 lakh to the members of the crew; 22 per cent between Rs 7 lakh and Rs 8 lakh; and the remaining 14 per cent between Rs 3 lakh and Rs 4 lakh per trip. If we take the weighted average, the payment comes to about Rs 7.60 lakh per trip. For six trips a year, the wages come to Rs 45 lakh per trawler.

TABLE 8.10
Percentage distribution of responses on average total running cost per trip of a trawler

Range of running cost before conversion (Rs in lakh)	Percentage response	Running cost after conversion (Rs lakh)	Percentage response
6–7	7.15	12	16.66
8–9	21.43	15	50.00
10–11	14.28	18	16.67
12–13	42.86	22.5	16.67
14–18	14.28		
	100.00		100.00

The total running costs per trip before and after the conversion, as envisaged by the trawler operators, are reported in Table 8.10.

The weighted average running cost before conversions, as calculated from Table 8.10, comes to Rs 11.20 lakh, of which wages for the crew represent Rs 7.60 lakh, as calculated before. That is, the cost of the crew alone constitutes about 68 per cent of the total running cost. The reported data does not give any clue as to any rise in crew cost due to diversification of a shrimp vessel by tuna longlining. But from a discussion with the trawler operators, we gather that if at all there is any rise it would only be marginal. Whatever the rise, the weighted average cost of running a trawler with tuna longlining would be Rs 16.25 lakh per trip. Assuming once again that a trawler makes six trips a year, the average running cost before diversification comes to about Rs 67 lakh, and after diversification to Rs 97 lakh or, say, Rs 100 lakh. Contrary to conventional marginal cost accounting, in fishing, these running costs, though apparently seeming to be variable, become fixed per year on the assumption of a certain minimum number of trips made in a year (in the present case it is taken as six). Other fixed overheads as estimated by Dixitulu, and also as gathered from the trawler companies, appear to vary between Rs 80 lakh and Rs 100 lakh per annum. Taking the mid-point of the range, the average of other fixed overheads comes to Rs 90 lakh. The total annual costs will, therefore, be Rs 190 lakh. At a sale price of Rs 5.16 lakh per ton, the cash-flow break-even point of the production of tunas comes to 36.82 tons.[10] In order to pay for the cost of capital required for diversification, and a reasonable return on the entrepreneur's invested capital (which shall include repayment of the instalment of the term loan), there should be sufficient amount of tuna catches beyond the break-even point, which we shall calculate now.

Dixitulu indicates that the capital cost for tuna diversification of the existing trawler would be Rs 48 lakh for 23–5 metres OAL and Rs 57 lakh for vessels of 28 metres OAL. His own calculations indicate that there would not be much of an increase in the percentage contribution (not more than 1 per cent) in the latter case, despite an additional investment of Rs 9 lakh. The preference for conversion should, therefore, be for the 23–5 metres vessel, at an approximate cost of Rs 48 lakh.

An analysis of the responses of trawler operators to our structured questionnaire reveals that the term loan requirement of the individual trawler operator varies between Rs 37.5 lakh and Rs 75 lakh for the proposed tuna

---

[10] We have included depreciation for calculating cash-flow break-even point owing to special nature of the asset, but for short-term policy decision it can be ignored.

longlining conversion. On an average, the term loan requirement comes to Rs 44 lakh. Assuming interest at the rate of 12 per cent per annum and a seven years' repayment period, the annual servicing cost of the loan (including repayment by half-yearly instalments) comes to Rs 9.48 lakh.

A further analysis of responses reveal that the working capital requirement for the trawler operators vary between Rs 15 lakh and Rs 40 lakh, depending upon the size of the trawler. The working capital requirement in the trawler industry primarily emanates from (i) variable expenses like fuel, fresh water, salaries of crew, and export transportation costs and (ii) repair and maintenance cost of the trawler. According to the expenditure statement provided by Dixitulu, the cost per trip-cycle (assuming 6 trips in a year) for item (i) comes to be Rs 16.83 lakh for the 23–5 metres vessel and Rs 21.20 lakh for the 28 metres vessel, and for cost item (ii) it is Rs 5 lakh and Rs 5.83 lakh, respectively. The total working capital requirement per trip-cycle, therefore, comes to Rs 21.83, say Rs 22 lakh, for a 23–5 metres vessel and Rs 27.03, or Rs 27 lakh, for the larger vessel. The average requirement of both types of vessel can be taken as Rs 25 lakh.

It is found from an analysis of the responses of the trawler operators that, on an average, they obtain advances from processors/exporters/other agencies to the tune of Rs 15 lakh. The advance per trip-cycle, therefore, comes to Rs 2.5 lakh.

The working capital gap for an average vessel type would, therefore, be Rs 25 lakh minus, 2.5 lakh, equal to Rs 22.5 lakh. The trawler-operators in their responses to the structured questionnaire said that they are prepared to provide, on an average, 20 per cent of the working capital requirement from their own sources, which comes to Rs 5 lakh. Hence, the requirement of loan for working capital finance will be Rs 17.5 lakh for a typical vessel. The servicing cost of this loan at the rate of 12 per cent per annum would be Rs 2.20 lakh, assuming it to be given in the form of overdraft.

The total servicing cost of both term loan and working capital loan will be Rs 9.48 lakh plus Rs 2.20 lakh, which equals to Rs 11.68 lakh. In order to absorb this servicing cost, additional catches of tunas should be 11.68/5.16 = 2.26 tons. The financial break-even point (FBE) of tuna catches, therefore, comes to 36.82 + 2.26 = 39.08 tons , say, 39 tons, or 6.5 tons per trip.

The financial break-even point of a diversified tuna fishing vessel appears to be high. It is also based on a price realization of Rs 5.16 lakh per ton. At

[11] Dixitulu, ibid.

a lower price realization, the FBE point will rise further. Even Dixitulu has remarked that the operations will be economical only when high value sashimi grade tuna or tuna loins are aimed at.[11] There may not be any constraint on the availability of oceanic tuna resources, as revealed from the analysis of their potential made earlier. But at the same time, we should keep in mind the migratory nature of tuna species which might lead to trip-failures. One or two such failures may turn the unit into a loss-making one. It is necessary, therefore, to go into further details on the economics of tuna diversification before the proposal is vetted by the MPDEA for financial closure.

It has already been mentioned earlier that banks/financial institutions are not too willing to touch any lending proposal from this sector. Once shy, it takes a lot of effort in confidence building. Keeping this in mind, we propose the following for the revival of the Deep-sea trawler industry.

RECOMMENDATIONS FOR THE FINANCIAL RESTRUCTURING OF DEEP-SEA FISHING

(a) All lending proposals for diversification should be vetted by the MPEDA in regard to the projections made and economics of the scheme, before being forwarded to banks/financial institutions (FIs).

(b) There should be a single agency for providing both term loan and working capital finance to the diversified company to have a total control of the operations.

(c) The loans made to a unit shall be counterguaranteed by at least four such trawling companies taking similar loans, who shall be jointly and severally responsible for repayment of loans together with interest. In other words, repayments of loans and interest thereof made to all the five units shall be the joint and several responsibility of all such units. In order to avoid litigation, banks/FIs should obtain an undertaking from all such loanees that in the event of any failure or shortage in periodical loan servicing of any borrower, the bank/FI would be entitled to recover the same from other borrowers, with equal proportions from the sale proceeds of the landed catches, the modalities of which are given in the subsequent paras.

(d) Repayment of term loans with interest and also the interest on overdraft should be linked to per trip-cycle of the individual trawler-vessel.

(e) Banks/FIs will inform the Marketing Association of the fishing harbour about the equated amount of repayments to be payable by each such trawler.

The Marketing Association, who shall be charged with the responsibility of issuing gate passes from the fishing harbour, will do so only on deposit of the equated amount of repayment as above, whether by the trawler-operator or by the buyer of the catches. The amount so recovered will be remitted by the Marketing Association to the concerned bank/FI. There shall be a tripartite agreement with the bank/FI, the trawler-operator, and the Marketing Association towards this effect.

(f) In order to enhance the credit-worthiness of the proposal and bring confidence in the minds of banks/FIs, the loans need to be collateralized. But the vessels, presently owned by the trawler-operators after a one-time-settlement, have lost their shiftability, and hence can no longer act as meaningful collaterals. It is necessary, therefore, to bring in new collaterals in the form of house-properties which most of the trawler-operators are reported to own. It is proposed that at least 50 per cent of the loan amount should be collaterized in such a manner. If banks/FIs are prepared to take further risk on trawler-venture there is no reason as to why the trawler-operators should also not risk their personal properties.

(g) Payment for diesel, which is a major component of working capital, should be made by the bank directly to the supplier.

(h) The rate of interest on such loans shall be 12 per cent or the prime lending rate (PLR) whichever is lower.

CONCLUSION

India being a peninsula and endowed with plentiful inland water resources has a large variety of fish stocks in both pelagic and demersal category. Commercial exploitation of Indian fishery began only in 1970s. The impetus came primarily from growing importance of shrimp in the export market. All attention was focussed on developing shrimp fishery. India soon emerged as the leading exporter of shrimp. In the following years Indian Seafood industry became predominantly shrimp oriented with cephalopoda playing the second fiddle. Not much attempt was made to diversify the industry nor there was any attempt to create a domestic base for the seafood export industry which would have enabled them to withstand shocks coming from the vagaries of world market, which is major cause of present day crisis of the industry. The high pressure on shrimp fisheries has caused overfishing of oceanic resources which made the industry turn towards aquaculture. Unfortunately, aquaculture in India, soon became synonymous to shrimp culture. It is time that Indian Seafood industry comes out of this shrimp-orientation and broad-base itself by diversifying to other fisheries. The Production Efficiency Index will enable them to develop a portfolio of fisheries for investment and market planning. At the national level, this Index can be used for allocation of developmental resources among different fisheries.

9

Disposition of Fish Landings in India and Export Performance

INTRODUCTION

The pattern of disposition of total fish landings in different forms reflects the changing tastes and demands of the consuming people. It also indicates to what extent the market is responsive to the changing need of consumers to obtain delivery of the same product in different forms. Although at a particular stage of socio-economic development the need of the domestic consumers may vary from that of the overseas consumers, ultimately with the integration of global markets and cross-penetration of culture some commonality in need orientation must evolve. In short, the demand of a section of overseas consumers to obtain fish products in a particular form may soon be the demand of a section of domestic consumers.

This chapter analyses the changing pattern of the disposition of total fish landings in India and the export performance. India appears to follow the global trend in disposition of total fish catches, as Table 9.1 shows.

FRESH AND FROZEN FISH

Table 9.1 shows that the share of the disposition of fish in fresh form has remained almost stationary, at around 65 per cent, during 1977–80; in the next four years the share increased to an average of 70 per cent, but in the following four years the share has declined to an average of 67 per cent. This alternate trend of up and down was, however, broken from 1990 onwards when the share began rising consistently, and by 1997 it has captured more than 70 per cent share of the total fish catches. The cumulative growth rate of fish in fresh form is found to be more than that of the total fish production.

Table 9.1
Disposition of fish catch in India

Year	Marketing fresh	Frozen	Cured	Canned	Reduced	Misc purposes	Offal for reduction	Total
1977	1517.6 (65.64)	159.6 (6.90)	497.0 (21.50)	4.9 (0.21)	91.6 (3.96)	31.5 (1.36)	9.7 (0.42)	2311.9
1978	1504.8 (65.25)	146.7 (6.36)	497.4 (21.57)	4.5 (0.20)	100.7 (4.37)	32.2 (1.40)	19.8 (0.86)	2306.1
1979	1510.6 (64.57)	151.6 (6.48)	512.7 (21.91)	4.3 (0.18)	120.0 (5.13)	33.3 (1.42)	7.1 (0.30)	2339.6
1980	1591.6 (65.17)	144.6 (5.92)	512.9 (21.00)	5.2 (0.21)	125.5 (5.14)	48.4 (1.98)	14.0 (0.57)	2442.2
1981	1668.0 (68.25)	118.1 (4.83)	441.7 (18.07)	5.5 (0.23)	162.0 (6.63)	27.2 (1.11)	21.4 (0.88)	2443.9
1982	1701.2 (71.87)	130.2 (5.50)	358.0 (15.12)	5.7 (0.24)	126.1 (5.33)	27.6 (1.17)	18.3 (0.77)	2367.1
1983	1784.6 (71.19)	145.0 (5.78)	423.9 (16.91)	7.3 (0.29)	110.6 (4.41)	22.4 (0.89)	12.9 (0.51)	2506.7
1984	1978.9 (69.14)	207.0 (7.23)	424.1 (14.82)	13.6 (0.48)	194.8 (6.81)	34.3 (1.20)	9.4 (0.33)	2862.1
1985	1843.3 (65.27)	196.4 (6.95)	562.2 (19.91)	9.8 (0.35)	170.0 (6.02)	28.6 (1.01)	14.0 (0.50)	2824.3
1986	1964.0 (67.22)	209.7 (7.18)	460.4 (15.76)	12.1 (0.41)	158.3 (5.42)	105.8 (3.62)	11.5 (0.39)	2921.8
1987	1968.3 (67.72)	179.0 (6.16)	526.4 (18.11)	4.9 (0.17)	190.6 (6.56)	19.0 (0.65)	18.4 (0.63)	2906.6
1988	2075.1 (67.32)	233.7 (7.58)	529.0 (17.16)	20.6 (0.67)	174.6 (5.66)	25.2 (0.82)	24.2 (0.79)	3082.4
1989	2301.4 (64.20)	261.2 (7.29)	590.9 (16.48)	28.7 (0.80)	315.2 (8.79)	62.0 (1.73)	25.1 (0.70)	3584.5

(Contd.)

Table 9.1 (contd.)

Year	Marketing fresh	Frozen	Cured	Canned	Reduced	Misc purposes	Offal for reduction	Total
1990	2497.2 (65.18)	285.7 (7.46)	598.8 (15.63)	29.3 (0.76)	322.3 (8.41)	63.4 (1.65)	34.8 (0.91)	3831.5
1991	2706.0 (66.91)	265.9 (6.58)	613.8 (15.18)	30.1 (0.74)	333.4 (8.24)	47.1 (1.16)	47.7 (1.18)	4044.0
1992	2798.4 (67.06)	284.8 (6.82)	590.2 (14.14)	25.9 (0.62)	355.8 (8.53)	47.4 (1.14)	70.6 (1.69)	4173.1
1993	3105.2 (68.31)	309.5 (6.81)	644.5 (14.18)	9.8 (0.22)	372.8 (8.20)	87.6 (1.93)	16.3 (0.36)	4545.7
1994	3247.6 (68.62)	310.3 (6.56)	651.9 (13.77)	12.2 (0.26)	397.3 (8.39)	74.3 (1.57)	39.3 (0.83)	4732.9
1995	3427.0 (69.85)	320.6 (6.53)	632.5 (12.89)	13.2 (0.27)	358.7 (7.31)	115.3 (2.35)	38.7 (0.80)	4906.0
1996	3727.0 (70.85)	397.9 (7.56)	658.4 (12.52)	11.0 (0.21)	314.0 (5.97)	114.5 (2.18)	37.4 (0.71)	5260.2

Cumulative growth rate (% p.a.)

	4.59	4.67	1.42	4.13	6.35	6.67	6.98	4.20

Source: *Hand Book on Fisheries Statistics,* Ministry of Agriculture, Fisheries Division, Govt. of India, New Delhi, 1996.

The rate of growth of frozen fish is close to that of fresh fish, and together they command 78.41 per cent of the total catches, but the share of frozen fish has not increased much during the past twenty years. This is despite the fact that the country now has about 376 freezing plants and 451 cold storages and, as revealed from the present study, many such plants suffer from considerable unutilized capacity. It appears that inspite of a rise in the world market for frozen fish, the Indian seafood industry has not been able to focus itself rightly in the world market. This includes both the domestic and export market. While we shall discuss the export market later it should be mentioned here that though the Indian seafood industry has developed considerable expertise at least in frozen shrimp, it has never given much thought towards the development of the domestic market for this product due to an untested hypothesis bordering on fantasy that Indian consumers are not capable of appreciating the product both in terms of quality and price. It is necessary for the industry to focus itself on the domestic market, not only for optimal utilization of freezing capacity but also to absorb the shocks emanating from the vagaries of the world market. The industry should make a detailed market research to explore the various possibilities in the domestic market.

CURED FORM

Cured fish products in India, which come under the FAO classification of Dried, Salted and Smoked group, have also followed the declining trend experienced worldwide. Its share in the total fish catches in India has fallen by 41.76 per cent between 1977 and 1996. The cumulative rate of growth is only 1.42 per cent per annum which is the lowest among all the groups of fish products.

CANNED FISH PRODUCTS

Canned products, which in India include specially processed value-added fish products, are particularly aimed at export markets. We have seen earlier that despite technological advancement in fish processing and the enthusiasm created for speciality fish products, the share of this product, which comes under the broad FAO classification of Prepared and Preserved fish products group, has not increased during the past twenty years. In India's total disposition of fish production, the share of canned items has also remained constant between 1977 and 1996, though during 1988–92 there has been a spurt in the export growth of this product.

REDUCTION

Fish oil reduction, particularly of sardines and sharks, and offal reduction for fish meal and manures are growing in importance in India during the past twenty years. Together they have a share of about 6.70 per cent of the total disposition of fish catches in India. But considering the volume of oil rich sardines and sharks captured in India the share is not very encouraging as compared to the world trade in the production of these reduced fish products.

EXPORT PERFORMANCE

INTEGRATED APPROACH

The Indian seafood export and its prospects cannot be viewed in isolation. It is an integrated function of domestic fish production, technological level of processing and preservation, a well-researched marketing focus and strategy both in respect of areas and product categories, and finally, the physical and financial carrying capacity of the exporters.

While analysing the export performance of advanced developing nations of Asia we have seen that this integrated approach has made them exploit fully the comparative advantages they enjoy. This is reflected in an increase in their share of the world market and better price realization. In many respects India's natural comparative advantages are superior to that of her neighbours but she is not able to realize the full potential of comparative advantages due to lack of proper marketing strategy supported by technical, managerial, and financial power. We have seen earlier that in the matter of price realization, India's performance is poor as compared to that of the neighbouring countries. Even in the case of crustaceans product, which is the principal component of India's export and with which the country began her export of marine products in the 1960s, the price realization is much lower as compared to many of the exporting countries of South East Asia.

There is a perception that though poor price realization of Indian seafoods as compared to other Asian competitors is a fact, it cannot be attributed to marketing strategy alone. To a large extent, there is a quality perception influencing prices of Indian seafoods because of the unhygienic practices followed by many processors. The unhealthy *inter se* competition between the exporters is also responsible for poor price realization. The presence of a large number of agents, merchant exporters with no stake in industry, erratic landing resulting in inconsistency in supplies, landings of small sizes in the absence of any scientific conservation, etc. are other factors leading to poor price realization.

It should be remembered that fish exports from a fish-eating country, like India, is always at the cost of some amount of domestic deprivation. The country may bear the cost of deprivation provided it gets back value in return, otherwise it amounts to a misallocation of resources.

We have already seen that given the existing trends in fish production, India will not be able to meet the domestic demand of fish in the next five years. But still, exports have to be made for the overwhelming national imperatives as we have discussed earlier. Under such compulsions, it becomes incumbent not only to increase the exports but also to obtain proper value for the country's natural resources.

The share of exports as a percentage of total fish production in India is on the rise, as will be evident from Table 9.2.

TABLE 9.2
India's production of fish and export of fish products

(in thousand tons)

Year	Total fish production (marine and inland)	Exports	Exports as percentage of fish products
1985–6	2876	83.65	2.91 (4.87
1986–7	2942	85.84	2.92 (5.01)
1987–8	2959	97.18	3.28 (5.86)
1988–9	3152	99.78	3.16 (5.49)
1989–90	3677	110.84	3.01 (4.87)
1990–1	3836	139.42	3.63 (6.06)
1991–2	4157	171.82	4.13 (7.02)
1992–3	4365	208.60	4.78 (8.10)
1993–4	4644	243.96	5.25 (9.21)
1994–5	4789	307.34	6.42 (11.42)
1995–6	4949	296.28	6.00 (10.94)

Cumulative growth rate (% p.a.)

	5.06	12.18	6.80 (7.63)

Note: Figures in brackets represent percentage to marine production. But these are not exactly representative because quite a few of the exportable species are products of aquaculture, e.g. shrimps.

Source: As in Table 9.1 and *Statistics of Marine Products Exports*, Marine Products Export Development Authority, Kochi, various issues.

From Table 9.2 it can be observed that the compounded rate of growth in exports is more than double the rate of growth of fish production. It is ue that the export of fish products does, to a large extent, depend upon domestic fish production, and hence one may be tempted to establish a statistical correlation between production and exports, which may be statistically valid, but in terms of the realities of the business (where re-exporting is a viable option) and imperatives of the country (containing the domestic consumption to increase exports and foreign exchange reserves), such a correlation may be meaningless. However, a back-of-the-envelope calculation suggests that at the present level of fish production and average export price realization, a 1 per cent rise in the share of export to total fish catches would fetch an additional export revenue of Rs 650 crore. But the right kind of solution may not lie in this arithmetic. In view of the growing demand for domestic consumption of fish, what is necessary is not to increase the share of export in total fish production (rather, there is a case for reduction) but to increase domestic production of fish and attempt towards higher price realization by developing an appropriate matrix of fish products for exports.

During the period 1987–8 to 1996–7 India's export of fish products has grown at a cumulative rate of 22.74 per cent per annum in rupee terms, but the growth is not as impressive as this when we calculate it in US dollar terms, and also by tonnage, as will be evident from Table 9.3.

TABLE 9.3

Comparative compounded growth rates of India's export of fish products and market expansion during 1996–7 by different modes of calculation

	Cumulative growth rate (% per annum)	Market expansion (%)
In US dollars	10.90	181.48
In Indian rupees	22.74	675.86
By quantity	14.56	289.18

It should be clear that there is a devaluation effect in the higher growth rate calculated in Indian rupees. Even during the recent time, say during 1993–6, Indian rupees got devalued against the US dollar by around 15.15 per cent or at a cumulative rate of 4.02 per cent per annum on an average. The same can be said of the market expansion of export in terms of rupees which apparently looks to be fantastic.

The rate of growth in quantitative terms gives a rather better indication of the export performance of the country, but when we compare it with the growth rate calculated in US dollars we find that it is lower than the former by 3.66 percentage points. This suggests that India is losing on the value of exports.

COMPETITIVENESS OF INDIAN EXPORTS

However, the market expansion of India's exports in US dollar terms is 181.48 per cent during 1987–8 to 1996–7, while the world export of fish products has expanded by about 128 per cent around the same period (1986–96). In terms of cumulative growth also, the country's performance is marginally better than the world average of 10.64 per cent per annum during 1985–96. The performance of these two growth parameters may make the country complacent if the market share parameter is not considered at the same time. India's average share of the world export was 2.13 per cent during 1975–84; it fell to 1.45 per cent during 1985–92; and, during the recent period (1993–6), it has risen marginally to 2.10 per cent. All these indicate a lack of competitiveness of the Indian seafood industry in the world export market. A comparative analysis of the market share performance of some of the neighbouring countries made in Table 9.4 will prove this point further.

TABLE 9.4

Comparative analysis of share of world fish exports of India and selected developing countries of Asia

Country	Average market share (%)		
	1975–84	*1985–92*	*1993–6*
China	2.16	2.88	4.88
Indonesia	1.47	2.20	3.30
Thailand	2.54	5.66	8.41
Taiwan	4.20	5.09	4.08
India	2.13	1.45	2.10

Source: Fishing Statistics: Commodities Year Book, Food and Agriculture Organisation, Rome, various issues.

Table 9.4 makes it clear that while for the other countries it is a case of marching forward in capturing the world export market of fish product, for India it is case of stagnancy or deceleration.

Tables 9.5–9.7 provide category-wise export data of fish productions from India by quantity, value (in rupees), and average unit price realization respectively from 1985–6 to 1996–7.

SHRIMPS VS. FISHES

From Table 9.5 we find that frozen shrimp, which in terms of quantitative share occupied the first place in India's total export in 1985–6, is presently holding the second position with a considerable loss of share—about 53.70 per cent. The first place has been captured by fish in fresh and frozen form, whose share of total exports of fish products has gone up by nearly 262.5 per cent during the past eleven years.

The export of frozen shrimp has grown at a cumulative rate of 6.35 per cent per annum representing 109.40 per cent expansion during the period under study. Both the growth parameters are, however, much lower than that of total fish exports. In contrast, export of fish in fresh and frozen form has not only grown at nearly double the growth rate of total exports, the expansion of the market is by 1538.15 per cent as compared to 352.12 per cent expansion of total export of fish products by India. It appears that India has begun following aggressively the world trend in export of fish in fresh and frozen form from the year 1990–1 onwards. But that does not justify the considerable deceleration in the rate of growth of export of shrimp products, particularly when this is the principal foreign exchange earner (more than 65 per cent) among all the seafood exports from India, as will be revealed from Table 9.6.

Among the frozen shrimps, the moderately value-added component in various peeled forms presently constitutes about 50 per cent of the total exports. While on the one hand, it is necessary to have a distinct policy direction towards increasing the production of shrimps, (particularly by aquaculture development, as we have indicated earlier) and on the other hand, by increasing the share of peeled forms for superior price realization for the same species. The country may be losing substantial value on nearly 50 per cent of her shrimp exports. This may be the reason for the lower price realization of India's crustaceans exports as compared to the other advanced developing nations of South East Asia. The issues involved may range from imposition of control over the average size of shrimp catches and their export to identifying appropriate matching of the type of value-addition by different shrimp species and their cost–benefit analysis. These issues need to be examined in greater depth to evolve a right kind of policy direction.

Such a policy direction must, however, precede the appropriate policy initiative towards upgradation of the existing freezing and processing capacity

TABLE 9.5
Category-wise export of fish products from India by quantity

(quantity in tons)

Year	Frozen shrimp	Frozen lobsters	Frozen cuttlefish/ fillets	Frozen squids	Fresh/frozen fish	Live items	Dried fish	Dried shrimps
1985–6	50,349 (60.18)	1650 (1.97)	5010 (5.99)	4619 (5.52)	10,561 (12.62)	–	8151 (9.74)	73 (0.10)
1986–7	49,203 (57.32)	1132 (1.32)	4694 (5.46)	9739 (11.35)	13,136 (15.30)	–	5368 (6.25)	19 (0.02)
1987–8	55,736 (57.35)	1863 (1.92)	9195 (9.46)	7621 (7.84)	14,904 (15.35)	–	5220 (5.37)	34 (0.03)
1988–9	56,835 (56.96)	1663 (1.67)	8262 (8.28)	16,374 (16.41)	11,234 (11.26)	–	3633 (3.64)	59 (0.06)
1989–0	57,819 (52.16)	2068 (1.86)	14,158 (12.77)	11,944 (10.78)	21,227 (19.15)	619 (0.56)	1081 (0.98)	3 (0.003)
1990–1	62,395 (44.75)	1600 (1.15)	11,596 (8.32)	16,667 (11.95)	42,340 (30.37)	655 (0.47)	1127 (0.81)	1 (0.0)
1991–2	76,151 (44.32)	1628 (0.95)	12,437 (7.25)	25,529 (14.85)	49,333 (28.71)	595 (0.35)	2136 (1.24)	59 (0.03)
1992–3	74,393 (35.66)	1613 (0.77)	18,951 (9.08)	30,364 (15.03)	75,376 (36.13)	573 (0.27)	3267 (1.57)	90 (0.04)
1993–4	86,541 (35.47)	1455 (0.60)	18,998 (7.79)	34,741 (14.24)	94,022 (38.54)	744 (0.30)	2602 (1.07)	–
1994–5	101,751 (33.11)	1251 (0.40)	28,145 (9.16)	37,194 (12.10)	122,529 (39.87)	1002 (0.33)	5850 (1.90)	235 (0.08)
1995–6	95,697 (32.30)	1677 (0.57)	33,845 (11.42)	45,025 (15.20)	100,093 (33.78)	1756 (0.59)	7367 (2.49)	358 (0.12)
1996–7	105,426 (27.88)	1312 (0.35)	31,778 (8.40)	40,924 (10.82)	173,005 (45.74)	2030 (0.54)	9372 (2.48)	1102 (0.29)
Cumulative growth rate (% p.a.)								
	6.35	(–)1.89	16.64	19.93	26.24	16.00	1.17	25.38

(Contd.)

Table 9.5 (contd.)

Year	Shark fins and Fish maws	Misc. items	Total
1985–6	231 (0.28)	3007 (3.60)	83,651
1986–7	237 (0.28)	2315 (2.70)	85,843
1987–8	273 (0.28)	2333 (2.40)	97,179
1988–9	315 (0.32)	1402 (1.40)	99,777
1989–90	295 (0.27)	1629 (1.47)	110,843
1990–1	152 (0.11)	2886 (2.07)	139,419
1991–2	317 (0.18)	3635 (2.12)	171,820
1992–3	316 (0.15)	3629 (1.74)	208,602
1993–4	–	4857 (1.99)	243,960
1994–5	–	9380 (3.05)	307,337
1995–6	–	10459 (3.53)	296,277
1996–7	–	13250 (3.50)	378,199
Cumulative Growth Rate (% p.a.)	3.99	13.15	13.40

Note: Figures in brackets represent percentage of total exports.
Source: Statistics of Marine Products Exports, The Marine Products Export Development Authority, Kochi, various issues.

Table 9.6
Category-wise export of fish products from India by value

(Rupees in lakhs)

Year	Frozen shrimp	Frozen lobsters	Frozen cuttlefish/fillets	Frozen squids	Fresh/frozen fish	Live items
1985–6	32,981 (82.87)	1445 (3.63)	1080 (2.71)	552 (1.39)	1715 (4.31)	—
1986–7	37,793 (82.04)	1432 (3.11)	1396 (3.03)	1727 (3.75)	2229 (4.84)	—
1987–8	42,578 (80.15)	2473 (4.66)	2231 (4.20)	1373 (2.58)	3023 (5.69)	—
1988–9	47,033 (78.68)	2360 (3.95)	2334 (3.90)	3809 (6.37)	2845 (4.75)	—
1989–90	46,331 (72.96)	3363 (5.30)	4731 (7.45)	2848 (4.48)	4820 (7.60)	134 (0.21)
1990–1	66,333 (74.25)	3430 (3.85)	4529 (5.08)	4499 (5.04)	9082 (10.17)	162 (0.18)
1991–2	97,912 (71.16)	5530 (4.02)	6091 (4.43)	10,938 (7.95)	14,320 (10.40)	198 (0.14)
1992–3	118,026 (66.78)	4334 (2.45)	11,888 (6.73)	15,190 (8.59)	23,241 (13.15)	349 (0.20)
1993–4	177,073 (70.73)	4268 (1.70)	13,818 (5.52)	19,247 (7.69)	29,599 (11.82)	671 (0.27)
1994–5	251,094 (70.23)	4603 (1.29)	22,401 (6.26)	24,510 (6.86)	44,657 (12.49)	1042 (0.30)
1995–6	235,643 (67.31)	5591 (1.60)	26,086 (7.45)	31,958 (9.12)	37,226 (10.63)	2130 (0.61)
1996–7	270,176 (65.56)	5488 (1.33)	27,237 (6.61)	29,045 (7.05)	63,692 (15.45)	3397 (0.82)
Cumulative growth rate (% p.a.)	19.16	11.76	30.86	39.13	35.15	49.80

(Contd.)

Table 9.6 (contd.)

Year	Dried fish	Dried shrimp	Shark fins and fish maws	Misc. items	Total
1985–6	761 (1.91)	6 (0.02)	312 (0.78)	946 (2.38)	39,798
1986–7	674 (1.46)	1 (0.00)	373 (0.81)	442 (0.96)	46,067
1987–8	663 (1.25)	5 (0.01)	482 (0.91)	292 (0.55)	53,120
1988–9	443 (0.75)	13 (0.02)	583 (0.98)	355 (0.60)	59,785
1989–90	147 (0.16)	–	599 (0.67)	526 (0.60)	63,499
1990–1	140 (0.10)	1 (0.00)	300 (0.21)	861 (0.63)	89,337
1991–2	288 (0.21)	26 (0.02)	896 (0.65)	1390 (1.02)	137,589
1992–3	480 (0.27)	21 (0.01)	1040 (0.59)	2174 (1.23)	176,743
1993–4	1627 (0.65)	–	–	4059 (1.62)	250,362
1994–5	2303 (0.64)	113 (0.03)	–	6804 (1.90)	357,527
1995–6	4471 (1.28)	176 (0.05)	–	6830 (1.95)	350,111
1996–7	4373 (1.06)	324 (0.08)	–	8404 (2.04)	412,136

Cumulative growth rate (% p.a.)

| | 15.69 | 39.43 | 16.24 | 19.96 | 21.51 |

Note: Figures in brackets represent percentage of total exports.
Source: As in Table 9.5.

Table 9.7

Table 9.7
Average price realization of category-wise export of fish products from India

(Rupees in lakhs)

Year	Frozen shrimp	Frozen lobsters	Frozen cuttlefish/ fillets	Frozen squids	Fresh/frozen fish	Live items shrimps	Dried fish	Dried shrimp
1985–6	0.66	0.87	0.21	0.12	0.16	–	0.093	0.082
1986–7	0.77	1.26	0.30	0.18	0.17	–	0.126	0.053
1987–8	0.76	1.33	0.24	0.18	0.20	–	0.127	0.147
1988–9	0.83	1.42	0.28	0.23	0.25	–	0.122	0.220
1989–90	0.80	1.63	0.33	0.24	0.23	0.22	0.136	–
1990–1	1.06	2.14	0.39	0.27	0.21	0.25	0.124	1.00
1991–2	1.29	3.40	0.49	0.43	0.29	0.33	0.135	0.441
1992–3	1.59	2.69	0.63	0.50	0.31	0.61	0.147	0.233
1993–4	2.05	2.93	0.73	0.55	0.31	0.90	0.625	–
1994–5	2.47	3.68	0.80	0.66	0.36	1.04	0.394	0.481
1995–6	2.46	3.33	0.77	0.71	0.37	1.21	0.607	0.492
1996–7	2.56	4.18	0.86	0.71	0.37	1.67	0.466	0.294
Cumulative growth rate (% p.a.)	11.96	13.97	12.47	15.97	7.24	28.84	14.37	11.23

(Contd.)

Table 9.7 (contd.)

Year	Shark fins and fish maws	Misc. items	Average price realization
1985–6	1.35	0.31	0.43
1986–7	1.57	0.19	0.51
1987–8	1.77	0.13	0.54
1988–9	1.85	0.25	0.61
1989–90	2.03	0.32	0.66
1990–1	1.97	0.30	0.77
1991–2	2.83	0.38	1.00
1992–3	3.29	0.60	1.06
1993–4	–	0.84	1.12
1994–5	–	0.73	1.18
1995–6	–	0.65	1.18
1996–7	–	0.63	1.30
Cumulative growth rate (% p.a.)	11.78	6.09	9.66

Source: As in Table 9.5.

of the seafood exporters to bring it in line with international standards, as parameters of hygiene requirements in shrimp products are comparatively stricter than other fish products, particularly in the markets of the developed world which claim the largest share of shrimp imports.

Lobsters

During the recent years, India appears to be exporting all the lobsters that she is capturing from her seas, whereas earlier, about 45 per cent were exported. Lobsters are among the highest bidders for price in the world fish market. For India also, the price realization is highest among all the fish products exported by the country. As such, lobsters' share of total fish exports is small; it is also exhibiting a decreasing trend, but their high unit value has caused overfishing. Presently, lobsters' export is fetching about Rs 55 crore for a quantity as small as 13 thousand tons. This is the economics behind their depletion. But if this situation is allowed to continue and no new lobsters stock is discovered then this species may become extinct. It is necessary, therefore, to conserve this valuable resource. An exploratory survey coupled with imposition of quota on the export of this fish species should be the strategy towards this direction.

Cuttlefishes and Squids

Frozen cuttlefishes and squids are among the traditional items of fish exports from India. Squids hold the third position, followed by cuttlefishes, in the Indian seafood exports. In quantitative terms (see Table 9.5) their share of the country's export has increased from 11.51 per cent in 1985–6 to 19.22 per cent in 1996–7. The percentage expansion of the market is 655 per cent during this period which is nearly double the expansion of total exports of the country in quantitative terms. This is also reflected in high cumulative growth rates of these two fish products.

In rupee value terms (see Table 9.6), the share of these two fish products is presently 13.66 per cent, increased from 4.10 per cent in 1985–6. The expansion percentage is 3348, that is more than 5 times of expansion in quantitative terms. This indicates a better price realization of these fish products in rupee terms. The rate of growth in unit price realization in rupees is 12.47 per cent for cuttlefishes and 15.97 per cent for squids. But in US dollar terms, the price realization is on the decline during the recent years. In fact, average price realization of almost all major fish exports from India is on the decline in US dollar terms as is evident from Table 9.8 .

Cuttlefishes are exported in different forms of which whole and whole cleaned form account for most of the share in quantitative terms. The share of the whole form is steadily on the rise and presently it is 55.34 per cent of

TABLE 9.8

Average unit value realization of major fish exports from India

(in US dollars per kg)

Fish items	1995	1996	1997	1998	1999
Block frozen shrimp	7.04	6.34	7.03	6.66	6.67
Frozen cultured shrimp	12.36	11.56	13.15	12.24	10.71
IQF shrimp	5.90	5.82	5.31	4.49	4.39
Frozen ribbon fish	0.65	0.66	0.72	0.60	0.57
Frozen whole cuttlefish	1068	1.93	1.87	1.46	1.40
Frozen whole cleaned cuttlefish	2.59	2.51	2.86	2.53	2.33
IQF cuttlefish	3.01	3.13	3.28	2.57	2.32
Frozen whole squid (baby)	1.68	1.39	1.41	1.60	1.51
Frozen whole cleaned squid	2.17	1.97	2.08	2.06	1.93
Frozen squid tube	2.30	1.81	1.91	1.87	2.05
Other frozen fish	0.78	0.81	0.90	0.89	1.12
Frozen whole cooked lobsters	10.83	10.24	10.30	8.29	10.95

Note: The twelve fish products reported here represent about 85 per cent of total fish exports from India.

Source: The Marine Products Development Authority, *Statistics of Marine Product Exports, 1999*, Kochi, 2001. The rupee values of average price realization are converted into US dollars using the periodic average exchange rate reported in, *Foreign Trade and Balance of Payments*, Centre for Monitoring Indian Economy, Mumbai, July 2001.

cuttlefish exports in quantitative terms but in value term the share is only 4.5 per cent. The low performance on the price front is due to the least unit price realization of this product. On the other hand, the whole cleaned form, which accounted for about 23.50 per cent of tonnage export of cuttlefishes in 1996–7, has claimed a 27 per cent share in value terms, because by a simple value addition (cleaning) it could claim a price of Rs 98,000 per ton as against the Rs 70,000 realized by the whole form. But unfortunately, sufficient attention is not being given to this rather simple value-addition which is reflected in the slow growth of the whole cleaned form in the export of cuttlefishes from the country.

The next value-added item is fillets, which command the second highest value among all the cuttlefish exports from India, followed by IQF (Individual Quick Freezing) form. But both the two forms require capital investments at a much higher level as compared to the whole cleaned form which may require little or no additional investments. At the same time, we should also remember that the demand of fillets is on the rise in the world market. It is necessary, therefore, to undertake a cost–benefit analysis to understand the economics of fillets before evolving a clear strategy for its development.

The Ink[1] of various selected cuttlefishes commands the highest unit value, around Rs 3 lakh per ton. It is the result of a value-finding exercise. Ink has usages in pharmaceutical industries. While discussing sharks, we have indicated that worldwide pharmaceutical usages of different parts of the shark are on the increase. The Indian seafood industry should keep its eyes and ears open to encash upon any such development to increase the value of exports.

The observations made above for cuttlefishes apply to the export of squids to a large extent. Price realization of squids in the whole form, which has a share of 36.88 per cent of tonnage export in 1996–7, is Rs 51,000 per ton, whereas the same in cleaned form, which has a share of 23 per cent, could claim a unit price of Rs 73,000 in the same year. For both cuttlefishes and squids the most important policy initiative should be towards replacing the export of the whole form by the whole cleaned form by imposing quota restriction on the former type, if necessary. Fillets have realized the highest unit value at Rs 2.06 lakh per ton, though they have a tonnage share of about 5 per cent in 1996–7. In value terms the share is about 15 per cent of the total squids export of the country.

FISH IN FRESH/FROZEN FORM

This fish product, which had a share of only 12.62 per cent of total tonnage exports of fish products from India in 1985–6, has apparently followed the world trend in the export growth of this product form. By 1996–7, it has increased the share to 45.74 per cent (see Table 9.5). The rate of growth is nearly double the rate of growth of total export of fish products from the country. The quantitative expansion between 1985–6 and 1996–7 is 1538.15 per cent, but in value terms (rupees) the expansion is 3613.82 per cent, which is also reflected in the very high cumulative growth rate of export of this product by value (see Table 9.6). All these indicate better price realization of this product in the export market in rupee terms. Virtually, price realization has more than doubled during the past twelve years.

India exports a large variety of fish in fresh/frozen form of which ribbon fish has accounted for 57.66 per cent of total tonnage in 1996–7, followed by mackerels (7.56 per cent) and pomfrets (7.35 per cent), seer fish (5.07 per cent), reef cod (2.18 per cent), and snappers (0.53 per cent). The remaining 19.65 per cent comprises sea breams, sea bass, croakers, tuna,

---

[1] This is a fluid in the mouth-sack of these fishes, which when thrown in the seawater, makes the water black, preventing the predator from seeing the fish.

oil-sardines, shark, fresh water fishes, eel etc. Some of these fishes are also exported in fillets form, such as reef cod, ghol, snappers, sea breams, emperor etc.[2]

RIBBON FISH

Although ribbon fish accounts for the largest share of tonnage turnover in the export of fish in fresh/frozen form and its contribution in rupee terms is also the highest at 38.03 per cent, the unit price realization is the smallest (Rs 24,000 per ton) among all the fishes coming under this group. During the past three years, there has not been much rise in the export price of this product from India in rupee terms despite a moderate devaluation of Indian rupees against US dollars. Since this fish constitutes the bulk of the Indian seafood export (more than 26 per cent of total exports, which is almost equal to that of shrimp exports), it is necessary to explore the possibility of adding some value, preferably labour-intensive, to this fish product.

Ribbon fishes are generally exported from India in whole, head-on form. Instances of export of headless form are rare. In some markets, such as Singapore, whole ribbon fish fetches better value than the cleaned ones. Here, there may be a case of value replacement instead of value addition. A 5 per cent addition to the export value of this fish product has the potential of adding about Rs 12 crore to the export earnings of the country.

POMFRETS

In contrast to ribbon fish, pomfrets, particularly the silver and Chinese variety, command the highest price in the export market of fish in fresh and frozen form. We have seen earlier that nearly 80 per cent of catches of pomfrets in India constitute silver and Chinese varieties, though in the latter type fluctuations in catches are very high bordering almost on uncertainty. We have also seen that pomfrets are reaching a stabilization state in their catches. Presently, India is exporting about 23 per cent of its total pomfret catches. Considering the stagnancy in production, it is unlikely that exports could be increased further from the present level, except by reducing their domestic consumption. The alternative strategy rests with moderate value addition keeping it within the broad frozen group.

TUNAS

Tunas together presently constitute a very small percentage of the total exports of fish products. The unit price realization is also very low—around Rs 35,000 per ton. Such a low price realization is due to the small size of

[2] *Marine Products Export Review, 1996–7,* The Marine Products Export Development Authority, Kochi, 1998.

tunas that are being captured in the coastal waters. We have indicated earlier that high value oceanic tunas remain virtually unexploited. It is expected that with the tuna longliners conversion of some of the shrimp trawlers, as proposed before, the picture will change.

MACKERELS

Mackerels have a tonnage share of about 7.56 per cent of fish exports in 1996–7. In value terms the share is about 6.20 per cent. We have noted earlier that mackerels suffer from high fluctuations in catches, which makes it difficult for commercial exploitation. But the fluctuations are not as high as those in oil-sardines. In Production Efficiency Indexes, Indian mackerels rank second. This calls for special attention to this fish species for the export market. The share of mackerels in total fish products exports from India is increasing, but very slowly. In respect of the unit price it is also very low, coming just ahead of ribbon fish. There remains considerable scope for exploiting this product for the export market. India does not have much of a presence in the expanding world trade in mackerels.

Eel is a new introduction to India's fish exports with a realised value of Rs 42,000 per ton.

Fillets are emerging as value-added items in the frozen fish group, which presently commands a price of Rs 1 lakh per ton.

All other fish items in this group together fetch an average unit price of Rs 48,000 per ton though their share of exports is small.

In conclusion, it can be said that though presently, fish in fresh/frozen form claims the largest share of the fish basket of India's export, its contribution is not matching in terms of value because of low unit price realization of most of the component-fishes. But there is considerable scope for its improvement by suitably modifying the production–export strategy. This requires the development of a separate product matrix for this group of fish products. In the short-run, India should target an additional value addition of at least 10 per cent, which may not call for much additional investments but the effect will be a rise in India's exports by about Rs 65 crore without changing the quantity base.

LIVE ITEMS

Live marine products are emerging as a good source of export revenue during the past eight years. Although their share in terms of tonnage, which has never gone beyond 0.60 per cent, is small (the world market is also not large), in terms of value they may soon be contributing at least 1 per cent to India's exports. The cumulative rate of growth of export of live items in value term is the highest (50 per cent) among all the fish product exports from India (see Table 9.6). The average price realization is of the order of Rs 1.70 lakh per ton (see Table 9.7).

The mud crab commands the major share (about 93 per cent) of the total quantity of exports of live items. The average price realized in 1996–7 is about Rs 1.20 lakh per ton. We have mentioned earlier that crabs are not receiving due attention despite their high potential in the export market, particularly the live mud crabs. Many countries have already gone in for aquaculture of the mud crab to exploit its potential for the export market. India should also follow the same course of action.

The grouper variety of live fish also fetches a good price in the export market (more than Rs 1 lakh per ton), but its production in India is small, as also its exports.

Live lobster fetches an average price of about Rs 8.5 lakh per ton, which is more than double the price realized in its frozen form. India's lobster resources are limited, they are also under depletion. The country should, therefore, explore the possibility of exporting lobsters more in live forms than in frozen form to increase the export earnings by the same quantity of exports.

India is presently exporting live items to the South East Asian markets. Europe has a vast market for live fish products. The lack of suitable facilities for air transportation and its prohibitive costs are acting as constraints for the Indian seafood industry to venture into European markets. We should point out here that the increase of export of fish in live form requires proper synchronization of different stages ranging from procurement in seas, maintenance of the stock in live condition, to their final export by air transportation. Unfortunately, such a synchronized system is yet to be fully developed in India. This may be the primary reason behind India not able to exploit the full potential of this market.

Dried Items

The share of dried fish products in the country's export has considerably decreased during the past 12 years both by tonnage and value, but the unit price realization has gone up from an average of Rs 9300 per ton in 1985–6 to Rs 46,600 per ton in 1996–7—the increase is more than 400 per cent. The Indian seafood industry has developed a mindset towards neglecting this fish product. This mindset might have been furbished due to a fall in its market in Asia (though it still commands more than 30 per cent of the world market). But the point that is missed is that in both America and Europe, the import of fish in dried, salted and smoked form is on the rise. There is no reason, therefore, why the Indian seafood industry would neglect this fish product. Instead, there is a case for increasing the export of this fish product. It is good that during the last three years there is a sign of improving upon the dismal performance of the earlier years.

We do not have export data of shark fins and fish maws for the full period. However, the limited data indicate that their share in the total fish exports is declining in quantitative terms (see Table 9.5), but in terms of unit price realization they rank second among all fish products exported from India.

In India, production of sharks is stagnating despite the fact that half of its potential remains unexploited. Most of the sharks produced are either consumed locally as a cheap fish item or simply wasted. Only a small amount finds its place in the export market besides its reduction in oil. We have mentioned earlier that shark fish products are growing in importance in the world market because of the increasing number of value-findings for different parts of this large fish. The Indian seafood industry should, therefore, gear itself up both for increasing the production of sharks and their exports.

STRATEGIC THRUSTS

We have already explained that an alternative marketing strategy for certain items, say lobsters, by value replacement, moderate value addition of certain fish items, say whole cleaned form instead of simply whole form, and labour-intensive value addition, in general, would increase the export earnings considerably, even for the existing tonnage level of exports. We have also mentioned that this requires the development of an appropriate product-matrix in order to make a concerted effort towards this direction.

We emphasize on moderate, labour-intensive value addition within the existing freezing and storage capacity of the country. The primary reason behind this strategic step is that it does not call for a large capital investment. As nearly 95 per cent of the Indian seafood industry are small-scale operators, this seems to be the appropriate strategy at least in the medium term. The production and export of high value-added fish products, like speciality items, should be limited to units having technical, marketing, and managerial skills and adequate net worth to venture into branded product transformation of fish commodities.

MARKET DESTINATION OF FISH EXPORTS

India's export markets for fish products are far more diversified now than what it was in the 1960s. During the first decade of India's export of fish products, the major markets were concentrated in Asia and America. Asia had a share of about 70 per cent of the total world exports in the 1960s in quantitative terms. Not only has this share come down to about 54 per cent in 1996–7, the composition of countries forming the Asian market has also undergone drastic changes. Ceylon (now renamed as Sri Lanka) was the leading country in the 1960s, claiming about 75 per cent share of India's export to the Asian markets and more than 50 per cent of the total exports of fish from Asia. The other countries in Asia were Burma, Singapore, and Hong Kong. Japan's share was negligible. The picture has undergone considerable changes over a period of time.

Things began changing in the 1970s with the 'discovery' of shrimp as a high value, exportable fish product and Japan emerging as its largest buyer. Shrimp virtually displaced all other fish products from India's export basket and Japan began pushing away all other Asian countries claiming a share of India's export. During this decade Japan's share of India's export of fish products was about 45 per cent by tonnage and about 63 per cent by value, which increased to about 56 per cent and 71 per cent, respectively, till 1982. But from then on the decline began and by 1996–7 the share had come down to 17.10 per cent by tonnage and 45.76 per cent by value, as will be evident from Table 9.9.

The fall in the share is due to the fact that besides shrimp, Japan does not have much of a share in the other fish products exports from India. The share of shrimp in the total export of fish has increased considerably during the last 15 years.

The United States, which had about 25 per cent of the share of India's export of fish, both by tonnage and value, during the 1960s began exhibiting a declining trend in the 1970s itself. By 1980 the share came down to an average of 15 per cent both by tonnage and value. There was no abatement of this declining trend in the following decade also and, by 1996–7, we find that in quantitative terms the share came down to 7.88 per cent. In value terms the fall is less drastic, as will be evident from Table 9.9.

Europe provides a rather contrasting picture. It had a share of only about 2.5 per cent in the 1960s but the share began rising continuously since then (though with some intermittent years of decline), and by 1996–7 we find that the share has increased to 18.82 per cent by tonnage and 19.17 per cent by value.

The expansion of markets in countries other than Japan, the United States, and Europe is remarkable. In the 1960s, India did not have much of a market outside these countries, but now she has market in almost all the countries of South East Asia and Oceania, a large number of countries in the Middle East, some countries in Africa and South America, and several small countries like Mauritius, Mali, Maldives, Malta, etc. All these 'other

TABLE 9.9
Changing pattern of the direction of India's export of fish products by major destination

Year	Japan			United States			European Union			Other countries		
	Share (%)		Unit price	Share (%)		Unit price	Share (%)		Unit price	Share (%)		Unit price
	Q	V	(Rs '000 per ton)	Q	V	(Rs '000 per ton)	Q	V	(Rs '000 per ton)	Q	V	(Rs '000 per ton)
1981–2	57.07	70.93	51	15.02	12.23	33	13.13	9.63	30	14.78	7.21	20
1982–3	52.62	72.68	64	15.70	11.78	35	9.48	7.57	37	22.20	7.97	17
1983–4	40.59	64.40	64	14.64	13.35	37	10.30	10.09	40	34.47	12.16	14
1984–5	48.19	67.75	63	15.84	14.71	41	10.32	9.03	39	25.65	8.50	15
1985–6	48.21	69.78	69	11.38	11.57	48	17.01	10.84	30	23.40	7.81	16
1986–7	43.45	67.42	83	13.22	12.22	50	20.82	12.90	33	22.53	7.46	18
1987–8	39.86	61.40	84	14.86	14.15	52	23.48	15.86	37	21.80	8.59	22
1988–9	35.89	59.69	100	13.56	11.73	52	33.90	20.82	37	16.65	7.76	28
1989–90	34.97	54.44	89	12.45	12.33	57	33.36	24.46	42	19.22	8.77	26
1990–1	27.32	51.30	120	11.59	12.23	68	30.82	25.96	54	30.27	10.51	22
1991–2	22.98	46.04	160	12.13	11.24	74	32.36	28.77	71	32.53	13.95	34
1992–3	19.73	45.34	194	9.64	10.78	95	32.33	28.92	76	38.30	14.96	33

(Contd.)

Table 9.9 (contd.)

Year	Japan			United States			European Union			Other countries		
	Share (%)		Unit price	Share (%)		Unit price	Share (%)		Unit price	Share (%)		Unit price
	Q	V	(Rs '000 per ton)	Q	V	(Rs '000 per ton)	Q	V	(Rs '000 per ton)	Q	V	(Rs '000 per ton)
1993–4	18.44	47.36	264	10.72	12.23	117	29.45	25.77	90	41.39	14.64	36
1994–5	17.41	45.98	307	10.45	13.71	153	23.17	20.31	102	48.97	20.00	47
1995–6	17.48	45.03	304	8.78	10.46	149	29.44	26.05	105	41.30	18.46	49
1996–7	17.10	45.76	292	7.88	10.58	146	18.82	19.17	111	56.20	24.49	47
Cumulative growth rate (% p.a.)	(–) 7.25	(–) 2.70	11.52	(–) 3.95	(–) 0.90	9.74	2.28	4.40	8.52	8.71	7.94	5.49

Source: Statistics of Marine Products Exports, various issues and Marine Products Export Review, various issues published by The Marine Products Export Development Authority, Kochi.

countries', which together had a share of 14.78 per cent by tonnage and 7.21 per cent by value in 1981–2, now claim a share of 56.20 per cent and 24.50 per cent, respectively.

JAPAN

Japan continues to be the major destination of India's shrimp production. In 1996–7 about 48 per cent of the total export of shrimp products was made to Japan alone. In value terms the percentage is about 63 per cent. Of the total import of fish by Japan from India, about 80 per cent by tonnage and 90 per cent by value constitute frozen shrimps.

India's share of Japan's total tonnage import of shrimps has also increased from 13.45 per cent in 1992 to 22.10 per cent in 1996. India holds the second position after Indonesia in Japan's import of frozen shrimps. As shrimps constitute the bulk of export to Japan, the average unit value realization from exports to Japan is also the highest among all countries.

Japan offers the largest market in fish products. She is the leading importer of all the four groups of fish products. The size of her import market is nearly US$ 16,800 million. The crustaceans product accounts for about 40 per cent of this market; fish in fresh, frozen or chilled from holds 43 per cent of the market; prepared and preserved form claims 14 per cent; and the dried form, the remaining 3 per cent. In this huge market of fish products, India has a share of only about 3.15 per cent in US dollar terms. If we exclude shrimps, in the remaining 60 per cent of fish imports of Japan, India's share is slightly more than 0.50 per cent. This is happening despite India having a long and well-established trade relationship with Japan through shrimp exports. India has not been able to make any inroads into the other fish markets of Japan. While diversification across countries is necessary for spreading the market risk, it is also necessary to see how much of a customer-country's total need could be satisfied so that intra-market risk is also sufficiently spread out. India is already a high-risk exporter in fish products because more than 65 per cent of her export earnings originate from shrimps and out of that, 63 per cent is coming from Japan alone. Any adverse development in the domestic production of shrimps or in the shrimp market of Japan could transmit severe shocks down the spine of the Indian seafood industry. The solution does not lie only in spatial diversification of the market. It is necessary to simultaneously develop a strategy for customer-oriented diversification to spread the intra-market risk. Such a strategy should not be limited to Japan only, but should be applicable in all the markets.

United States

Another destination of India's fish products is the United States. This case is similar to that of Japan in many respects. Shrimps account for more than 75 per cent of the United States' import of fish products from India. Japan and the United States together control more than 75 per cent of India's shrimp exports. This makes the India's seafood industry highly vulnerable as we have discussed above. While making the country analyses, we have mentioned that fish consumption in the United States is showing signs of decline and it is likely that the United States might turn out to be a net exporter of this fish product. India should draw up an appropriate strategy to face the emerging situation.

The average price realization from export of all fish products to United States is much lower than that from Japan, as will be evident from Table 9.9. The primary reason behind this is that the unit price realized on shrimps in the United States is about 38 per cent lower than what is obtained in Japan.

European Union

In the European Union, the United Kingdom is the largest importer of fish products from India followed by Belgium, the Netherlands, Italy, Spain, and France. These six countries together account for about 80 per cent of the import of fish products from India by tonnage and more than 84 per cent by value. Of these, shrimps constitute about 47 per cent by tonnage and more than 66 per cent by value. The other countries of the European Union, namely Germany, Portugal, Greece, Denmark, Switzerland, Ireland, Norway, etc., presently have a small share each. While making the country analyses, it was shown earlier that Italy provides a much larger scope in fish marketing, and some of the presently smaller importing countries who are net importers also provide a similar scope. India should evolve an appropriate strategy to develop these markets for diversification of exports.

South East Asia

In South East Asia, China is the leading importer of Indian fish products, particularly frozen fin fish, which accounts for more than 90 per cent of the total fish imports from India. We have indicated earlier that China, which is hitherto a net exporting country in fish products, might turn out to be a net importer because of her rising level of domestic consumption. Although the Indian seafood industry is presently facing certain problems in dealing with this country, as is revealed in this study, it is hoped that with the liberalization of the Chinese economy these problems would be ironed out and India would be able to increase her exports to China.

After China, Hong Kong is the second largest destination of India's fish exports in South East Asia, followed by Malaysia, Singapore, Thailand, Korea (R), and Taiwan. We have seen earlier that all these countries are net exporters of fish products. Many of them are re-exporters. In a competitive market economy there is nothing wrong in using advanced marketing network and the brand name of a competitive country/exporter to channelize the products of the not so advanced country/exporter. But this can only be a short-term strategy. In the long run, a country/exporter has to develop its own network and brand name for moving towards direct marketing. China is already engaged in such a process. The moving out strategy needs careful planning as otherwise, it may unsettle the existing trade relations and goodwill of the country/exporter due to retaliatory moves by the competing countries.

Africa and the Middle East

The oil-rich countries of the Middle East and Africa have always remained good markets for fish products, but India has risen to the occasion only recently. India's export to the Middle East countries is increasing slowly. In 1996–7 the share was 2.56 per cent by tonnage and 0.015 per cent by value, which has increased to 4.57 per cent and 3.08 per cent respectively in 1997–8. India is yet to break into the oil-rich African countries, such as Nigeria whose consumption of fish is on the rise. These are the markets which need to be cultivated by India. Hitherto, these markets are fed by the United States, Canada, and European countries. Hence, India is expected to face stiff competition in exploiting these markets.

ANALYSIS OF TOTAL FISH PRODUCTS EXPORT PERFORMANCE OF INDIA

India's export performance in fish products during the past 19 years could be divided into two periods. The first period (1980–1 to 1988–9) is marked by consistent but slow growth, while the second period (1989–90 to 1998–9) is marked with rapid growth followed by stagnation, as is revealed by Table 9.10.

PROBLEM IDENTIFICATION AND ACTION PLAN

The cumulative rate of growth in quantitative term in the second (recent) period is more than three times the growth rate in the first period, while in value terms it is double the rate of growth of the first period. The reason behind the variation in growth rates in quantitative and value terms is the low average price realization, as discussed earlier. On the one hand, relatively

TABLE 9.10
Export of total fish products from India

Year	Quantity (in thousand tons)	Value (Rupees in crores)
1980–1	(–) 75.6	(–) 234.84
1981–2	70.1 (–5.5)	286.01 (21.79)
1982–3	78.2 (11.55)	361.36 (26.35)
1983–4	92.7 (18.54)	373.02 (3.22)
1984–5	86.2 (–7.01)	384.29 (3.02)
1985–6	83.7 (–2.90)	398.00 (3.57)
1986–7	85.8 (2.51)	460.67 (15.75)
1987–8	97.2 (13.29)	531.20 (15.31)
1988–9	99.8 (2.60)	597.85 (12.55)
1989–90	110.2 (10.42)	635.00 (6.21)
1990–1	139.4 (26.49)	893.37 (40.69)
1991–2	171.8 (23.24)	1375.89 (54.01)
1992–3	208.6 (21.42)	1767.43 (28.46)
1993–4	244.0 (16.97)	2503.62 (41.65)
1994–5	307.3 (25.94)	3575.27 (42.80)
1995–6	296.3 (–3.58)	3501.11 (–2.07)
1996–7	378.2 (26.64)	4121.36 (17.72)
1997–8	385.8 (2.00)	4697.48 (13.98)
1998–9	302.9 (–21.48)	4626.87 (–1.50)

Cumulative growth rate (% per annum)

(a) 1980–1 to 1988–9	3.13	10.94
(b) 1989–90 to 1998–9	10.64	22.71
(c) Coefficient of Variation		
1988–9 to 1996–7	35.71%	61.00%

Note: Figures in brackets indicate percentage rate of growth.
Source: As in Table 9.9.

low value fish items have begun filling the fish export basket of India, which at one time was dominated by shrimps, and, on the other, the commodity export culture prevalent in the Indian seafood industry led to ignoring the necessity of adding at least a moderate value to a sizeable bulk of fish exports. These two together have contributed to the low average price realization of India's fish exports.

It should be pointed out here that mere freezing does not add value to the fish products. It only adds costs both to the producer and consumer (thawing). In effect, a frozen fish takes away value from the live or fresh fish. At the same time, we must agree that freezing is an essential condition in the fish trade. Hence, value addition must take place before the freezing operation. Any amount of saving of labour of the user (consumer) of the fish is considered as value addition by the producer. Every time a producer puts his fish in the freezer he should stop and think whether he has added value even to the smallest extent, like simple cleaning and gutting of the fish. If he has not, then some one else will do it and take away a piece of his cake.

The trend projections of India's export of fish products that will be made in the following section can be altered substantially by changing the mindset of the Indian seafood industry.

PROJECTIONS

For the purpose of developing appropriate trend equations we have relied on the export performance data of the entire period, and also for the second period, namely 1989–90 to 1998–9. The trend equation found by regressing the data for the whole period has revealed a comparatively poor significance level. Export data are taken by value for the purpose of projection.

The trend equations so developed are given in Table 9.11 with other regression parameters.

TABLE 9.11
Trend equations

Eq. no. Year	Constant(a)	(b) Coefficient	SE(b)	t. value	R^2	Level of significance
(1) 1989–90 to 1998–9	17.89	500.34	30.63	23.83	0.97	1%
(2) 1980–81 to 1998–99	(–) 986.27	263.49	29.32	12.55	0.83	1%

Projections of exports made on the basis of equation (1) can be regarded as highly probable (optimistic) and that by equation (2) as less probable (pessimistic). These projections are given in Table 9.12.

The Indian seafood export industry is presently in the grip of severe sickness, with half of its net worth eroded. If the present state is allowed to continue it is unlikely that India's export of fish products will go behond the pessimistic estimates. In Chapter 10 the causes of this sickness are analysed and measures for its turnaround proposed.

TABLE 9.12
Projection of fish product export from India

(Rupees in crore)

Year	Optimistic	Pessimistic	Average
2001	6022	4547	5285
2002	6522	4811	5667
003	7023	5074	6048
2004	7523	5338	6431
2005	8023	5601	6812

CONCLUSION

We have earlier seen that compared to many fish exporting countries of South East Asia the price realization in fish exports from India is much lower. Although our study has also revealed that the existing trend in fish production in India will not be able to cope with the rising domestic demand, there is an imperative to earn precious foreign exchange by exporting fish products. The strategy should, therefore, be not so much to increase the share of export in our total fish production but to attempt towards a higher price realization by developing an appropriate matrix of fish products for exports. There exists considerable scope for improving the export earnings by suitably modifying the production–export strategy of fish in fresh, chilled and frozen form.

It is true that during the past decade India has tried to diversify its export markets for fish products but the progress is slow. Still, more than 65 per cent of India's export earnings come from shrimps, of which Japan alone imports 63 per cent. This overwhelming dependence on a single product and a single market makes the country a high-risk exporter. Any adverse movement either in domestic production or in the shrimp market of Japan may pose a serious threat to the seafood export industry of India. It is, therefore, imperative for India to step up the pace in diversification of both product-range and product-markets.

10

The Crisis

INTRODUCTION

Presently the Indian seafood export industry is passing through a period of severe crisis. During 1994–6 several things started happening one after the other, and some even simultaneously, which adversely affected the industry. During this period the United States Food and Drug Administration (USFDA) began imposing restrictions and embargoes from time to time on the import of shrimps from India. This was followed by rejection of a large number of consignments of shrimps, particularly in cooked form, by the USFDA on grounds of poor quality. To add fuel to the fire the Federal Court of the United States, by a judgement on a case relating to endangered species, prohibited the import of seafood from countries who were not using the Turtle Excluder Device (TED) in their fishing nets (India was one such country at that time).

The United States was followed by major European importers in imposing restrictions and a ban on Indian seafood imports. The lead was taken by Italy, Spain, and France. Italy and Spain rejected a number of consignments for alleged salmonella contamination. Some of these consignments were also not returned but destroyed by the sanitary authorities.

In Japan (the largest importer of shrimps from India), the market became sluggish, and due to recession the Yen appreciated against the US dollar. The aftermath of the disastrous Kobe earthquake also dampened the general demand for imports.

Back home, the shrimp aquaculture, which is the major source of shrimp exports, came under the spell of widespread disease, resulting in a drastic fall in production.

All the above happenings, though not very uncommon for any food export industry, put the seafood export industry to severe losses, and almost the entire industry became sick due to a low level of net worth, from which it is yet to come around. The following sections attempt to analyse the basic problems of the industry and then suggest measures towards their resolution.

FINANCIAL ANALYSIS

Although the crisis was already brewing in 1994, the financial impact of the crisis and its severity were observable first only in 1995. Though published data of the financials of the marine foods industry indicate the onset of the crisis in no uncertain terms, the severity of the crisis was found to be more pronounced at the unit level when we conducted the survey. The official data of most of these units often suppressed the real impact of this crisis for the reasons explained later. However, many of these units have come out with the real picture in response to our questionnaire under assurance of anonymity.

First, we present an analysis of the financial aggregates and ratios relating to the marine foods industry published by the Centre for Monitoring Indian Economy (CMIE) in 1999. It should be pointed out beforehand that the sample used by CMIE is not quite amenable to rigorous statistical testing. However, the sample includes almost all the top companies in the marine foods industry. There are also problems of discontinuity of the reporting firms, and of data aggregation. In order to circumvent some of these problems we have normalized the data wherever possible.

Our findings are divided into three broad heads, namely, profitability and net worth, assets (capacity) utilization, and working capital.

PROFITABILITY AND NET WORTH

Various profitability and net worth ratios are presented in Table 10.1 for the years 1991–2 to 1997–8.

From Table 10.1, it can be observed that the growth in sales was arrested in 1995–6 when it came to be negative for the first time during the first five years under study. It was marginally positive in 1996–7 (due to a lower base in the earlier year), but plummeted down by more than 21 per cent in 1997–8. The operating profit ratio, which by nature should be more or less constant or somewhat upward over a period, came down by more than 50 per cent in 1995–6 and was a substantial negative in 1997–8.[1] This may be due to high, unabsorbed overheads or poor sales realization or both.

---

[1] For a detailed discussion of financial ratios and their interpretation see, Bhattacharya, Hrishikes, *Total Management by Ratios: An Integrated Approach*, Sage Publication India (P) Ltd., New Delhi, 1995.

TABLE 10.1

Profitability and net worth ratios of the Indian marine foods industry

	1991–2	1992–3	1993–4	1994–5	1995–6	1996–7	1997–8
Growth in sales (%)	76.9	50.3	103.3	70.3	(5.6)	3.1	(21.3)
Operating profit (% of net sales)	6.6	8.0	5.1	8.7	3.9	3.4	(4.5)
Interest (% of sales)	4.40	4.07	6.94	5.51	9.52	10.60	12.45
Profit after tax (% of sales)	4.6	3.9	2.9	4.1	(4.4)	(5.3)	(14.8)
Growth in net worth (%)	–	96.97	10.78	37.5	(10.71)	(8.37)	(33.33)

Source: Financial Aggregates and Ratios, Centre For Monitoring Indian Economy, Mumbai, 1999.

Worse is the situation with the Net Profit Ratio (PAT/Sales) which continued to be negative from 1995–6 onward, and in 1997–8 the negative was nearly 15 per cent. The reason behind this is that due to a substantial fall in the operating profit, loan burdens of the marine food industry continued to rise, primarily because of non-repayment and the consequent cumulative effect on interest accruals. (There had also been contracting of ad-hoc loans to stave off the crisis). The resultant rise in interest burden on the industry is reflected in the interest ratio which increased from 5.51 per cent in 1994–5 to 12.45 per cent in 1997–8—a rise of more than 127 per cent in three years.

When the industry is suffering from a continuous fall in profit the net effect is felt in the erosion of net worth. The fact that there has really been substantial erosion of net worth is indicated by the growth in net worth for 1997–8 in Table 10.1. The percentage fall in net worth is continuous and quite substantial during the past three years under study. We may recall that the CMIE sample includes almost all the top companies in the marine foods industry which are presumed to have been driven by good marketing strategy, superior quality management, and product diversification. Despite that, erosion of net worth had taken place, which accentuated the crisis.

Calculated on the base of 1994–5, erosion of net worth during the following three years is close to 46 per cent. If we go by the official definition of industrial sickness and use erosion of 50 per cent of net worth as the criteria for determining sickness, then the entire marine foods industry should be declared as sick.[2] It may be pointed out here that the actual erosion of net worth is much more than what is reported. Our study has revealed that many firms continued to show profit in their books of accounts when, in fact, they were making losses, as otherwise they feared that banks would stop renewing or enhancing their loan limits. Any policy formulation for the revival of the seafoods industry should, therefore, take into account this hidden erosion of net worth.

Assets (Capacity) Utilization

Three key assets utilization ratios as given in Table 10.2 are used here.

Table 10.2

Assets (capacity) utilization ratios of the Indian marine foods industry

	1991–2	1992–3	1993–4	1994–5	1995–6	1996–7	1997–8
VOP/Gross fixed assets (times)	8.13	2.77	2.47	2.39	1.61	1.49	1.05
VOP/Net fixed assets (times)	8.92	2.87	2.58	2.61	1.78	1.67	1.24
VOP/Total assets (times)	2.73	1.07	1.05	1.07	0.77	0.70	0.53

Note: VOP = Value of production
Source: As in Table 10.1.

Substantial underutilization of assets is clearly evident from 1995–6 onwards. The utilization of assets was highest during 1991–2; in reality there had been overutilization due to a spurt in sales, which on the other hand, fuelled uncontrolled capacity creation in the following years that remained unutilized to a large extent. Utilization of fixed assets was more or less stabilized at around 2.5 times during 1992–3 to 1994–5, down from more than 8 times registered in 1991–2. During the last three years (1995–6 to 1997–8), the ratio continued to exhibit a steep fall and hovered around 1 in the last year. Such a low level of fixed assets utilization has serious consequence on the profitability. When a capacity is created, financial burdens are contracted simultaneously in terms of maintenance and up-keep, and loan servicing. When capacity cannot be utilized for sales

---

[2] The Sick Industries Companies (Special provisions) Act 1985 defines sickness as where at the end of any accounting year peak net worth of a unit is eroded by way of its accounting losses of that year being equal to or more than 50 per cent of its peak net worth in the immediately preceeding five accounting years.

generation, the financial burdens continue. This eats into the profitability of the enterprise, as seen for the marine foods industry in the earlier paragraphs.

WORKING CAPITAL

The continuous fall in profitability, with its resultant impact on erosion of net worth ultimately caused a severe cash crisis in the marine foods industry. If the level of net worth of the units were high, the crisis would not have been so severe. In an industry, which is highly risk prone, both in respect of production and sales (export), the level of net worth should match with the riskiness to make it sustainable, particularly when it does not have much of a domestic base to neutralize the shocks. But in the Indian marine foods industry—95 per cent of which are in the small scale sector[3]—the average level of net worth is hardly more than 5 per cent of sales. This is the reason why we find that the industry, which boasted about being a sunrise industry even in the early 1990s, just crumbled under pressure, be it environmental, hygiene regulations, or the adverse movement of the overseas markets. The industry is yet to recover from the shocks because not much of additional net worth is being created. The consequences are felt in the worsening of the working capital position of the industry, as will be revealed from Table 10.3.

TABLE 10.3
Working capital ratios of the Indian marine foods industry

	1991–2	1992–3	1993–4	1994–5	1995–6	1996–7	1997–8
Current ratio	1.27	1.46	1.22	1.38	1.24	1.25	1.01
Average days of finished goods(on cost of sales)	43	43	54	52	76	86	92
Average days of debtors (on sales)	19	26	22	23	35	38	40
Average days of creditors (on purchases)	14	20	37	40	55	63	102
Average days of current liabilities (on sales)	22	50	69	65	70	78	106
Growth in current liabilities (%)	–	72.22	34.56	18.49	(10.69)	–	(1.93)

Source: As in Table 10.1.

[3] Presently, the capacity utilization is barely 25 per cent of installed capacity. See, Editorial, *Seafood Export Journal*, December 1996, p. 3.

From Table 10.3 it can be seen that holding of two major current assets, namely finished goods inventory and debtors have increased considerably during the last three years—close to double by 1997–8 as compared to 1994–5. During the same period, creditors increased by more than double and total current liabilities, which includes bank finance for working capital, also went up by about 63 per cent. The resultant effect is felt in the worsening of the current ratio which was just about 1 in 1997–8. This indicates that the industry is presently operating with zero net working capital. The real situation is much worse than what is revealed from the above analysis based on published accounts. While debtors could be real, as these are mostly export related and also the current liabilities, which include bank loan for working capital, the same may not be true of finished goods inventory. We have explained later that in reality, there might not be much of a stock except in the stock statement submitted to banks, as otherwise, there would be no drawing power.

We may also notice from Table 10.3 that there was not much of a growth in current liabilities during 1995–6 –1997–8, and the level remained stable at its position of 1995–6. When there is a fall in sales, fall in profitability, and erosion of net worth, stability of the current liabilities only indicate the virtual dependence of the industry on current liabilities as the only available source of finance to draw sustenance. No repayment is also made because that will dry up the only available source. The problem was further accentuated when a part of this fund was diverted towards upgradation of fixed assets, which shall be discussed later.

We shall now analyse the sources of problems of the Indian seafood industry and our micro-level findings delineating the reasons behind such problems. The micro-level findings are based on responses to our questionnaires, discussions with a large cross-section of processors/exporters, and also banks. We had developed four separate sets of questionnaires meant for (i) banks/FIs, (ii) insurance companies/ECGC, (iii)processors/exporters, and (iv) trawler operators. All these questionnaires were pre-tested for validity before launching the survey. The respondents to questionnaires were guaranteed absolute anonymity in order to elicit truthful information.

As the universe for each category was rather small, we did not draw any sample, but went in for a population survey.

RISK FACTORS

Fish products export of India suffers from high fluctuations. The level of fluctuations is higher by value than by quantity. The coefficient of variation (CV) is 35.71 per cent by quantity and 61 per cent by value of exports for the period 1989–90 to 1998–9. These high fluctuations make the seafood export business more risky as compared to other manufacturing industries engaged in exports. Part of this risk also emanates from fluctuation in domestic catches and procurement of fish. During 1989–97 CVs of marine and inland procurements are found to be 8.51 per cent and 16.85 per cent, respectively, and for the total catches, it is 11.58 per cent. The low level of fluctuations (CVs) during the recent period is due to stagnancy of Indian fish production that we have seen before. Whatever may be the reason it appears that at present, riskiness in fish production is rather low.

All these fluctuations are to a large extent uncontrollable, and hence should be considered as an overriding condition of the industry. Analysis of the responses to the structured questionnaires sent to banks and financial institutions on this aspect indicate that riskiness varies between moderate to high. Of the total respondents, 46.67 per cent consider the Indian seafood industry as highly risky while the remaining 53.33 per cent consider it as moderately risky. None of the respondents were willing to put the industry in the low risk category. The rankings of major risk factors done by banks/FIs are given in Table 10.4

From Table 10.4, it may be noticed that the major risks to the Indian seafood industry are emanating from the overseas markets in the form of fluctuations in demand and in prices, and the imposition of a ban on imports. All these factors ultimately affect the sales and thereby the net worth via losses. Banks and FIs do not rank high the domestic risk factors, like fluctuations in catches or in domestic prices of fish. The Indian seafood industry should, therefore, take measures to protect itself against the riskiness inherent in the industry and, thereby also provide comfort to the financing agencies.

In the ultimate analysis, it is the sales (export) which is at risk due to varying conditions of the business. Any measure to build protection against the sales-risk must, therefore, be linked to sales (export).[4] Sales-risk is a combination of fluctuations in production (catches) and fluctuations in sales (export). Risks, as measured by CVs, are 11.58 per cent and 61 per cent, respectively, for production (by quantity) and exports, (by value). Assigning a weight of 40 per cent to production and 60 per cent to exports, the weighted average riskiness of the Indian seafood export business comes to be 41.23 or, say, 40 per cent. The usage of the Coefficient of Variation as a measure of the sales-risk has been explained before. We should point out here that fluctuation in production (catches) is not restricted to fluctuation in volume alone; it is also related to the unit size of catches of different

[4] For further insight in the matter, see, Bhattacharya, Hrishikes, *Banking Strategy, Credit Appraisal and Lending Decisions*, Oxford University Press, New Delhi, 1997.

TABLE 10.4

Banks/financial institutions' ranking of major risk factors of the seafood industry

(per cent of response)

Risk factors	Bank							
	1	2	3	4	5	6	7	8
Fluctuations in international demand	20.00	13.33	26.67	6.67	13.33	13.33	6.67	–
Fluctuations in international price	13.34	40.00	13.33	33.33	–	–	–	–
Fluctuations in supply (catches)	–	13.33	–	26.67	20.00	13.34	13.33	13.33
Fluctuations in domestic prices of fish	–	–	–	–	13.34	6.67	26.66	53.33
Ban on imports by foreign countries	60.00	20.00	20.00	–	–	–	–	–
Exchange rate fluctuations	–	–	6.67	6.67	13.33	40.00	20.00	13.33
Change in government policy	–	20.00	–	13.33	20.00	13.33	33.34	–
Sudden outbreak of disease	20.00	6.67	20.00	6.67	20.00	–	13.33	13.33

exportable species. As the latter could not be quantified in exact terms, though diminution of the unit size during recent time is reported by the respondents, we have attempted to address this problem by assigning a somewhat larger weight to the CV of fish production.

One straightforward way of protecting the seafood export industry against the sales-risk is to provide for net worth equivalent to about 40 per cent of the average sales of a unit. But the recommendation of such a high level of net worth cannot stand the test of feasibility, firstly because the industry at its present level of distress is not able to bring in any significant level of net worth, not to speak of the required 40 per cent. Secondly, this high net worth level would result in such a low leverage in the capital structure of the firm that the return on equity would no longer be attractive to an entrepreneur.

But at the same time, we must re-emphasize that net worth is the ultimate solution to the management of inherent riskiness of a business. As mentioned before, this is more true for a high-risk business like the seafood industry. An alternative way of reducing the pressure on net worth is to create a domestic base for the seafood industry which can absorb a part of the shock emanating from the vagaries of the world market. We, therefore, suggest that the Indian seafood industry should be encouraged to create a domestic sales base to the tune of 25 to 30 per cent of sales, including exports. In other words, a unit intending to make an export sales of Rs 100 lakh must ensure at least a domestic sales of Rs 33.33 lakh. For a domestic sales level of 30 per cent, the value of domestic sales will be Rs 42.86 lakh.

The above may require the building up of adequate infrastructural support, including the establishment of a cold chain, that is a series of cold storages placed along the delivery line of the product. But one need not wait for that. A beginning can be made by using the existing channels of distribution to the domestic market. Waiting for the infrastructure to develop and then do business is not quite entrepreneurial. It is the development of the market which forces the infrastructure to develop and, it is not always necessary and desirable that the government should build up this infrastructure; the seafood industry itself can do it.

The remaining part of the risk should be covered by building up the net worth of the firms engaged in seafood exports.

As a part of the study, we have called from banks/financial institutions details of the financial performance data of at least the two best performing firms of the seafood industry in their portfolio. Analysis of these data indicates, among other things, that the units which have been able to withstand the crisis plaguing the seafood industry during the past three years have net worth/sales ratio varying between a range of 22.60

per cent and 55.56 per cent (though, the number of such units is around 3 per cent of the total). None of these export-oriented units has a domestic sales base.

We have shown before that during the past 4–5 years there has been continuous erosion of net worth in the Indian seafood industry due to all the major risk factors, as appearing in Table 10.4, operating against the industry simultaneously with various levels of intensity. The worst hit are the small scale sector which comprises 95 per cent of the Indian seafood industry. As a result, loans made to these firms by banks/FIs became non-performing assets. A ranking of the major causes for such borrowal accounts turning non-performing, as ascribed by banks/FIs, is given in Table 10.5.

TABLE 10.5
Ranking of major causes ascribed by banks/financial institutions behind a borrowing account of a seafood industry becoming an NPA

(per cent of responses)

Causes	Bank						
	1	*2*	*3*	*4*	*5*	*6*	*7*
Inadequate net worth	28.57	28.57	21.43	14.28	7.15	–	–
Inadequate or negative NWC	28.57	21.43	28.57	14.28	7.15	–	–
Inexperience of borrower	7.14	28.57	21.43	21.43	14.28	7.14	–
Diversion of funds	35.72	14.29	21.43	14.28	14.28	–	–
Under insurance of stock	–	–	–	7.14	–	92.86	–
Greed	21.43	42.86	21.43	14.28	–	–	–

It can be observed from Table 10.5 that besides diversion of fund (to which we shall come later), inadequate net worth or net working capital (NWC) (which follows from the former) rank very high as a contributory cause towards a seafood borrowal account turning NPA. This is because when risk factors start operating against a unit it is net worth which, in the final analysis, can absorb the shocks that come one after the other or simultaneously. But the seafood industry, particularly the small scale sector, already had a low net worth which got eroded to a large extent or even wiped out in some cases due to the prevailing crisis.

In accordance with the above findings and the risk level calculations made by us we suggest that seafood export firms, besides having a domestic sales-base as mentioned above, should have a net worth level of 10 to 15 per cent of export sales.

DIVERSION OF FUNDS

Our study has revealed that there has been substantial diversion of funds from working capital to fixed assets (the amount of such diversion could not be estimated for obvious reasons). Reasons for such diversions, as gathered from a cross-section of the industry are, primarily, the European Union (EU) ban and the imposition of stringent hygiene requirements by the importing countries of the developed world, which necessitated quick upgradation of processing technology. In the absence of net worth and banks' unwillingness to lend further, a part of the working capital loan from commercial bank was diverted towards upgradation of technology. It is also likely that some amount of working capital fund was also drawn out by the entrepreneurs as subsistence remuneration.

As indicated before, in order to ensure that banks do not cut back on their working capital limits, many such units continue to show in their books of accounts and stock statements the existence of stocks which are not there (or not exportable in declared value due to the EU ban, changing market, etc.) and profits in their Profit and Loss Account which are never earned. This syndrome has caused further erosion of their already dwindling net worth and forced them further into the debt-trap from which they are unable to come out. As a consequence, almost all loans made by banks to such units lost serviceability and became non-performing, as mentioned before.

In Table 10.5 bankers have given a high rank to 'Greed' as one of the major causes for a borrowal account turning NPA. This particularly refers to the aquaculture of shrimps. From a discussion with a cross-section of bankers it is revealed that many of the shrimp aquaculturists are not experienced in proper pond management and the risk they run in the absence of it. Hence, greed plays the dominant role. Many of them do not follow the standard per square metre concentration of shrimps, as more concentration gives a bumper profit. And that is the primary incentive for taking a high risk by beating the standards. As a result, oxygen supply in the pond is reduced and water environment is polluted, leading to disease and harvest failure. Bankers ascribe this as the primary reason for an aquaculture borrowal account turning bad. While it is difficult to control the greed of a businessman, the solution to this problem lies in multiculture practices that we have discussed earlier.

METHODS OF LENDING AND NORMS FOR CURRENT ASSETS HOLDING

Our study has revealed that presently, among banks, there does not exist any uniform methods of lending to the seafood industry. Previous to the Reserve Bank of India's withdrawal of the directive regarding the MPBF (Maximum Permissible Bank Finance) system, banks in general were following the Second Method of Lending for working capital finance to units belonging to the seafood industry. After such withdrawal of directive, 66.67 per cent of the respondent banks are found to continue with the Second Method of Lending and the remaining 33.33 per cent are reported to have devised their own method of lending.[5]

There is also a wide variation among banks on the current assets holding norms for the seafood industry as will be revealed from Table 10.6.

One bank has reported that it follows an aggregate current assets holding norm as 45 per cent of projected sales or average percentage of current assets on actual sales during the past 3 years. Another bank has said that there is no fixed norm. Holding is decided on case by case basis considering the peculiar feature of a unit.

The wide divergence in the practices of the commercial banks, as revealed in Table 10.6 and in the above paragraph, makes it difficult to calculate any average current assets holding norms for the seafood industry. For the same reason it is also not possible to make any comparative study with the norm applicable to similar other sectors like agriculture, food processing, etc.

However, about 86 per cent of responding banks agreed fully to the existence of the following peculiar needs of the seafood industry in deciding norms for current assets holding. The agreement level of the remaining 14 per cent varied between 50 per cent and 83.33 per cent.

(a)   The need to pay substantial amount of cash advances to fishermen, vessel operators, agents, aquaculturists etc;

(b)   The need to maintain high inventories owing to uncertain export market conditions;

(c)   The need to build up raw materials and semi-processed stocks at the time of landing seasons to enable the exporter to supply round the year;

(d)   The need to hold stocks in overseas markets to make this industry competitive in supplying value-added consumer packs;

(e)   The need to maintain low value, slow moving stocks due to the

[5] For a further discussion on the MPBF System and Methods of Lending, See, Bhattacharya, Hrishikes, Ibid.

TABLE 10.6

Banks' norms for current asset holdings for the seafood industry

(per cent of responses)

Name of current asset	Weeks holding											
	0	1	2	3	4	5	6	7	8	12	16	24
Raw materials	33.33	22.22	11.11	–	22.22	–	–	–	11.12	–	–	–
Work-in-process	36.36	9.10	9.09	–	9.09	–	–	–	27.27	–	9.09	–
Finished goods	–	–	14.30	–	28.57	14.28	–	–	–	42.85	–	–
Sundry debtors	11.11	11.11	–	–	11.11	–	22.23	–	22.22	11.11	–	11.11
Advances	42.86	14.28	14.29	–	14.28	14.29	–	–	–	–	–	–

compulsion in buying inspite of landing fluctuations in the availability of exportable species;

(f)   The need to operate without confirmed orders or Letter of Credit due to the inability of the processor to control the size/grade of material available and the product mix, especially when using raw materials from the sea.

When asked whether in view of the above peculiar needs of the seafood industry there is any need to change the existing norms, 53.33 per cent of the respondent banks answered in the affirmative; 40 per cent in the negative; and the remaining 6.67 per cent were undecided.

Overall, the above findings indicate that there is an overwhelming need to evolve new current assets holding norms for the purpose of working capital finance. For this purpose we have made in-depth study of the working of the plants of processor–exporters, the physical holdability of finished fish products in freezing conditions, and market practices between exporter and importer. Our findings and suggestions for improvement are discussed later.

INSURANCE AND ECGC COVERAGE

During our discussions with bankers and exporters a lot of misgivings were expressed about claim settlement by the insurance companies and Export Credit Guarantee Corporation (ECGC). In order to understand the actual position on this aspect we sent a structured questionnaire both to the general insurance companies of India, the ECGC, and the banks. We have not received any response from the ECGC but two general insurance companies and almost all the banks who have commitments in the seafood industry have responded.

TABLE 10.7

Distribution of insurance claim settlement by insurance companies

(per cent of total responses)

Range of settlement of claim submitted (per cent)	Response percentage	No. of cases (average) (per cent)
100 – 80 of claim amount	14.28	10
79 – 60 of claim amount	42.87	70
59 – 40 of claim amount	14.28	15
39 – 20 of claim amount	28.57	5
	100.00	

On the claim settlement by the insurance companies the responses of the banks are summarized in Table 10.7.

The responses of the insurance companies on this aspect are given in Table 10.8.

TABLE 10.8

Claim settlement by insurance companies

	1992–3	1993–4	1994–5	1995–6	1996–7	1997–8	Average
Percentage of claims settled on numbers received	97.47	68.15	62.47	95.85	60.26	100	80.70
Percentage of claims settled on amount of claims made	132.01	43.46	25.69	117.55	142.46	119.66	96.80

Note: According to one insurance company 'claims received during the year' should be read as 'Incurred claims' and 'claims settled' during the year should be read as 'claims paid during the year' which include unsettled claims lodged in the earlier years as well as the claims lodged during that particular year.

A comparison of Table 10.8 with Table 10.7, reveals that on an average claims settlement reported by banks match with that of insurance companies. Misgivings about the claim settlement may be restricted to about 20 per cent of the cases.

The claim-settlement experience of banks with ECGC is somewhat poorer than that with insurance companies. Fifty per cent of respondent-banks reported claim settlement between 90 per cent and 80 per cent and the remaining 50 per cent reported settlement between 79 and 60 per cent of the amount claimed.

Considering all these we propose a single agency and a comprehensive insurance policy which shall also include credit risk. The rate of premium should be inclusive of premiums hitherto being paid for both the general insurance and ECGC cover.

Although the premium for the ECGC cover is paid by the exporter (the bank debits the borrower's account), on the non-receipt of an export bill financed by a bank the claim amount received by the bank from the ECGC goes towards adjusting the loan amount. But the borrower is not freed from the liability to the ECGC, and the bank having received the claim amount now acts as an agent of the ECGC to recover the amount from the exporter–borrower. That is, though the credit risk is taken by both the bank and the exporter–borrower and the premium for export credit risk coverage is paid by the latter only, in the event of the credit risk maturing the lender is paid off but the liability of the exporter–borrower is not extinguished. The position appears to be anomalous and inequitable. While all the exporter–borrowers we have met are resentful about it we wanted to find out from the banks their reactions to the problem. In response to our structured question whether banks desire the ECGC guarantee cover to be with recourse or without recourse to the borrower–exporter, 26.67 per cent of the respondents say that it should be without recourse to the borrower; 60 per cent are in favour of continuing the existing with recourse system; while 13.33 per cent are undecided.

It is heartening to note that at least a quarter of the respondent banks feel the inequitability of the existing system. We strongly recommend that the comprehensive insurance-cum-credit risk policy as suggested above should be without recourse to the borrower–exporter. The 'moral hazard' problem can be taken care of by limiting the settlement payment to 80 per cent of the claim amount.[6]

NPA SETTLEMENT

Banks'/FIs' anxiety over recovery of NPA accounts is understandable because even when an account is secured by assets/collaterals the existing legal process is time consuming. Given the situation, a one-time lumpsum settlement is often found to be beneficial. As we shall be proposing such an adjustment procedure, we wanted to find out from banks about the average time taken for the recovery of a NPA account through the normal legal process. Banks' responses are summarized in Table 10.9.

TABLE 10.9
Time taken for recovery of NPA through normal legal process by banks/FIs

(per cent of responses)

Time taken	Small loans including (SSI)	Large loans
Beyond 10 years	22.22	36.36
Less than 10 years but more than 9 years	11.12	27.28
5–8 years	44.44	18.18
Less than 5 years but more than 3 years	22.22	18.18
1–3 years	–	–

[6] The 'moral hazard' problem is characteristic of the insurance business, where the insured party has little or no incentive to recover the loss or defaulted amount when the entire amount is payable by the insurance company.

It may be observed that according to 77.78 per cent of the respondent-banks, NPA accounts under small loan category are not realized through legal process in less than five years time; in the case of large loan the percentage is 81.82. None of the respondents has ticked the recovery time as 1–3 years. If this is the average experience of banks/FIs in the matter of recovery through the normal legal process, it is advisable to work out a compromise formula within the Net Present Value framework.

PLAN FOR FINANCIAL RESTRUCTURING OF THE SEAFOOD EXPORT INDUSTRY

Keeping in view the problems faced by the Indian seafood export industry as discussed so far and looking for a long-term strategy for its revival we propose the restructuring of the industry as follows:

INTEREST RATE RATIONALISATION

(a) Considering the risk parameters revealed in the study both in catches and exports we have proposed earlier that export oriented units should have a domestic market-base equal to 25 per cent of sales. This will enable these units to absorb shocks from the fluctuations in catches and vagaries of international market.

As these units will remain dominantly export oriented and as domestic sales are aimed at supporting the export activity, the loans required for this domestic operations should carry interest rate applicable for packing credit.

(b) There should be a uniform rate of interest applicable to seafood exporters both for term loans and working capital loans as the acquisition of fixed assets and stocks are both directed towards export of fish products.

(c) Inland water fishery units who make supplies to processor/exporters (considered as deemed exports) should similarly enjoy a rate of interest on bank/institutional loans equal to that applicable for exporters.

(d) The seafood export industry earn foreign exchange which are precious for building up the country's foreign exchange reserve. Although in terms of monetary conversion, 100 US dollars may equal to Rs 4600 in the exporter's hand, this is not so equal at the country's foreign chests. In the absence of foreign exchange reserve, the country may have to borrow at least at, say, 2 per cent over the London Interbank Offer Rate (LIBOR) (which is around 5.85 per cent now) in the international market. The foreign exchange earners should share this benefit they are bringing to the country by a reduction in the domestic interest rates on the loans taken from banks/FIs which will closely equalize the domestic rate with the international market rates. If we go by LIBOR, exchange earners should at least get a minimum of 6 per cent reduction in the domestic rate of interest, that is, the prime lending rate (PLR). RBI's export Refinance Scheme for banks should take care of this.

FUND FOR RECONSTRUCTION

(a) Earlier we have suggested that in order to cover the sales-risk the Indian seafood industry should, besides having a domestic sales base to the tune of 25 to 30 per cent of sales (including exports), also have a net worth equivalent to 10–15 per cent of sales. Therefore, at a minimum average export level of Rs 5000 crore per annum during the next five years, the industry immediately requires an infusion of net worth to the tune of Rs 500 crore.

But we have already shown that most of the units having suffered from erosion of net worth may not be able to bring in further capital to the business. Hence, the fund should come from an external source.

Keeping the above in mind we propose a Reconstruction Fund of Rs 500 crore to be made available to the seafood industry by the central government not by way of grant but by way of loan, repayable over a period of 5 years, the modalities of which are discussed below.

(b) The central government, directly or through an approved agency, shall raise this fund from the market by issuing 5-year tax-free bonds aggregating Rs 500 crore, carrying an interest of 8 per cent per annum. At the current rate of income tax, the before tax return on such bonds comes to 11.94 per cent per annum, which is an acceptable return in the financial market. The purpose behind the issuance of tax free bonds is to enhance its marketability.

Interest payable on such bonds will be Rs 40 crore annually or Rs 200 crore in five years' time. The aggregate of principal payment and interest will thus be Rs 700 crore.[7]

The issue may be named as the Marine Products Export Development Bond.

(c) The central government will impose an additional cess of 2 per cent for 5 years on all exports of seafood and accumulate the proceeds in a separate Repayment Fund. The aggregate amount of the fund along with interest income outstanding at the end of 5 years will be utilized to

---

[7] If however, it is felt that issuance of tax free bonds interferes with the competitive spirit of the market, the bonds can as well be issued without any tax benefit on interest. At the current rate of income tax, the equivalent nominal rate on the bonds will be about 12 per cent per annum which is quite attractive considering the present interest rate scenario. The net liability of the government to the public would, however, remain the same due to tax recovery on interest payments.

reimburse the government for repayment of the principal amount of the bonds along with interest aggregating Rs 700 crore.

We have found from Table 9.11 that average seafood exports from India during the next 5 years will be around Rs 6000 crore per annum at constant price level and exchange rate of 1998–9. A 2 per cent cess on such exports will fetch an amount of Rs 120 crore per annum or Rs 10 crore per month. This amount will be continuously invested in approved government securities or in a scheduled bank with a minimum targeted return of 11.25 per cent per annum. The value of the funds thus invested will become Rs 800 crore at the end of 5 years on monthly compounding basis, which is more than sufficient to reimburse the government towards repayment of the principal amount of Reconstruction Bonds along with interest. If we now consider the rising prices of fish products in the global market and moderate devaluation of rupees, India's export will be more at current prices and hence, the collection of cess amount. We, therefore, visualize that aggregate value of the fund will be more than Rs 800 crore at the end of 5 years. Any amount left in this fund after making full reimbursement to the government should be transferred to the seafood Exporters Assurance Fund as envisaged later in para (a).

While making this recommendation we are aware that the seafood export industry is already subjected to a MPDEA cess of 0.3 per cent and Agriculture Product cess of 0.5 per cent. Added to this is 0.2 per cent deduction imposed by the Export Inspection Agency, making the total to 1 per cent. The additional cess proposed here would raise the total deduction to 3 per cent. A question may arise whether the industry is able and willing to bear the total burden. For this purpose we held detailed discussions with the Executive Committee members of the Seafood Exporters Association of India (SEAI) and also directly with a cross-section of exporters at various places of the country. The consensus opinion that emerged from these discussions was that although this would create some hardships, considering the long-term benefit of the industry they were willing to bear the burden. Besides, as mentioned in Chapter 7, the additional cess will be absorbed by the raw materials market under the given price equation. This additional cess will in fact, benefit the industry in the long-run by lowering down the procurement price. When this additional cess is withdrawn after five years it will enter as profit element in the above equation because, by 5 years, the market will get stabilized at a lower price level.

(d) The amount of Rs 500 crore so raised by the government by issuance of bonds will be made available to the seafood export industry immediately for boosting up their net worth. Allocation of this amount among various units of the seafood industry will be made in proportion to the average exports of a unit for three years ending with 1996 multiplied by a growth factor of 1.18 (derived from second regression equation in Table 9.12) to take care of the projected exports for the next two years, beginning 1 April 2000. This formula is applicable for those units which are presently operational, registered before 31 March 1996, and/or intend to continue business in the future.

For units which are registered after 31 March 1996, and still in operation, average exports will be calculated based on their annual performance till 31 March 2000. The derived figure should also be similarly multiplied by the above factor for the purpose of allocation of the Reconstruction Fund.

For those units which are not operational or who do not intend to continue in the business, availability of the fund will be calculated without using the growth factor.

Utilization of the Reconstruction Fund by individual recipients, both operational and non-operational, will be discussed in subsequent paragraphs.

(e) A business undertakes business risk and a bank or financial institution undertakes lending or credit risk. If these risks mature into reality the capital (net worth) of the business or the bank absorbs the loss. As in other businesses, in the seafood industry also, the losses have eroded the net worth of the businesses. For the banks, the erosion of net worth is by a fraction of bad loans. While in all businesses, capital is always at risk along with other liability holders who directly or indirectly participate in the business, in the case of banks the depositors who supply about 90 per cent of the liability fund to the bank cannot or do not participate in any manner in the lending business of the banks. Hence, the deposit fund has to be insulated against the losses that might occur due to lending risks taken by banks.

Capital loss due to risk taken by both the borrower and lender is a common factor but not the loss to the depositors. The seafood industry must, therefore, repay in full the deposit fund lost in them along with its servicing cost. We, therefore, propose that 90 per cent of the outstanding loans due to a bank or financial institution should be repaid at the rate of average cost of deposit/fund of a bank or financial institution. The following methodology is proposed to make the recommendation operational.

(f) The principal amount of loan will be crystallized on the date when the account is declared non-performing for the first time. Interest will be charged on 90 per cent of this principal amount at the rate equal to the average cost of deposit of a particular bank less the refinance differential already received by the bank from the Reserve Bank of India on export finance. The total amount payable as on an appointed date will thus be calculated. (The appointed or cut-off date should be decided in consultation with the Indian Banks Association, the Marine Products Export Development Authority (MPEDA), and Seafood Exporters Association of India.)

Assuming further that banks have an interest spread of 4 per cent and that it takes a minimum of 5 years' time to recover a loan amount in default in the normal legal process, the amount so payable on the appointed date should be discounted further by a factor of 0.822 as one time settlement payment.

(g) The Reconstruction Fund available from the government as stated above may be first utilized towards one time settlement of banks' claim as above. Or, if a unit so desires, it can pay up 50 per cent of the settled amount and carry the remaining 50 per cent repayable during the next five years at the normal export finance rate (refinanced). The bank will recover this amount by making a deduction from the proceeds of the export bills. The rate of such deduction will be decided by the respective bank depending upon the amount repayable and the projected export receivables.

As an alternative to the last proposal and also for future assurance to the banks for repayment of loans, we suggest that the Authority administering the Duty Exemption (Pass Book) Scheme (DEPBS) shall send the entitlement of an individual exporter–borrower directly to the named banks. DEPB is freely tradeable (and mostly, it is so) as a cash item. We propose to route the trade through the lending banker. If repayment instalment of a loan is due, the bank will deduct the sum from the proceeds of DEPB sale through the bank.

In the normal course also, DEPB entitlement (sale) can be linked with repayment of term loan instalments.

(h) The Reconstruction Fund and the Repayment Fund will be administered by the MPEDA. The Reconstruction Fund will be made available to a unit only through a designated bank who will have the first right to adjust the overdue amount calculated as above as one-time settlement. The balance, if any, will be credited to the current account of the firm.

In the event of the reconstruction fund received by a unit being less than the one-time settlement amount, the remaining debit balance will be converted into a term loan repayable over a period of five years in the manner described above.

For units which are closed or non-operational and which do not intend to continue in the business, the amount of reconstruction fund will be first utilized towards adjustment of the one-time settlement amount. The remaining amount, if any, will be remitted back to the administering agency, namely the MPEDA. If the one-time settlement amount is more than the reconstruction fund so received the balance amount will have to be absorbed by the bank. Such cases are likely to be very few.

While making these recommendations we have kept in mind the hidden erosion of net worth that had taken place behind the published accounts of various units in the marine food industry. This is owing to their anxiety to show profits (when in fact there were losses), as otherwise they feared that banks would not renew or enhance their loan limits. Therefore, our recommendation in respect of net worth infusion to individual units from the Reconstruction Fund avoids any such calculation of erosion, real or otherwise. It aims at building up the net worth of the industry by linking it to sales. If the hidden erosion of net worth has been funded by bank loans (which is likely in most cases), then by making repayment to the bank the real net worth position of the unit is restored.

In the case of units where no real erosion of net worth has taken place (such cases may be very few) or their present net worth position is more than 10 per cent of sales, they should also get similar infusion of additional net worth calculated in the same manner as stated above. The reason behind this is that the firms which have performed better inspite of the present adversities should be rewarded by allowing them to perform much better in the future. In the case of the second group of firms, the same reason applies, though their present net worth could be net of erosion that had taken place in the earlier years. This approach is also equitable in the sense that these groups of firms are also required to contribute towards repayment of Reconstruction Bonds by a cess of 2 per cent on all their future exports.

(i) We suggest that even after full repayment of Reconstruction Fund the additional cess on exports be continued, though at a lower level of 1 per cent, to develop a Fund to protect the industry from natural or market catastrophes. This Fund may be named as seafood Exporters Assurance Fund. This fund should be administered by a Standing Committee of the MPEDA with its Chairman acting as Chairman of this Committee. The terms of reference of this Standing Committee and the rules for administering and utilization of the Fund shall be formulated in consultation with the government and the industry.

TERM LOANS

(a) As the seafood industry is already suffering from excess capacity in freezing and chilling processes, banks should not encourage the creation of any further capacity in these areas. However, term loans for upgradation of technology to conform to international standards will be provided by banks/financial institutions. The rate of interest will be equal to the rate chargeable on pre/post-shipment credit, as mentioned before. The quantum of such loan will be equal to the cost of upgrading the existing capacity with due synchronization.

(b) Finance for upgradation or increasing capacity in the high value-added segment of prepared and preserved fish products (speciality items) should be provided by banks to units having comparatively high net worth and adequate technical and managerial skills to produce value-added fish products. A minimum debt–equity ratio of 1.5 is proposed for such units.

NORMS FOR WORKING CAPITAL FINANCE

We should emphasize that the Reconstruction Fund being made available to the seafood export industry is one part of the revival package for the industry. The other part is making provisions for the availability of term loan and working capital finance to the industry from banks. The two together form the total package. An analysis of published financial data of the Indian seafood industry has revealed that presently, the industry is operating with zero net working capital. The real situation may be worse than this, as explained in this book.

It is true that the Indian seafood industry has suffered losses primarily due to its inability to absorb shocks. The inexperience of some units, and lack of proper management methods in which many of them are not trained, might also have contributed to the crisis. But the experience they have gathered, though at a cost to themselves and also to the banks, is one that no amount of training could have provided. All this will be lost if they are not allowed to continue in the business. This requires support from banks in providing adequate working capital finance.

Our study has revealed that presently, there is a wide variation among banks on the required level of holding of current assets by the seafood export industry. Virtually, there does not exist any uniform norm. This has also made it difficult for us to undertake any comparative study with norms applicable to similar other sectors such as agriculture, food processing, etc. we have, therefore, made unit level study of a cross-section of different segments of the Indian seafood industry to evolve certain general norms.

Our suggestions are as follows:

(a) We are not in favour of imposing any margin requirement of working capital loan during the reconstruction period of the seafood industry. The so called margin, hitherto provided by the units, are found to be on paper only as they do not have adequate net worth to actually provide the required margin. We propose to bring both the pre-shipment and post-shipment credit at par and hence, in the former case also the loan amount will be based on 100 per cent of projected cost of procurement/production.

(b) The study has revealed that a well procured and produced stock of fish can be carried safely under appropriate freezing condition for a period of not less than 6 months without any loss of value. We, therefore, propose that the outer limit of the holding norms for such finished stocks should be 6 months. This is necessary to improve the bargaining power of the exporters for better price realizatio. in the export market as indicated in the study.

(c) Banks should inclua book-debts, such as advance payment to suppliers of raw materials, fo the purpose of assessment and financing of working capital. It has been observed that on an average three weeks' procurement cost of materials rotates throughout the year as advance payment to suppliers. We, therefore, propose such a norm for this particular current asset.

(d) In the seafood industry, raw materials (fish) need to be processed immediately before storage. In other words, raw materials immediately enter into work-in-process on arrival. The holding of materials on work-in-process depends upon the value chain of individual units. Hence, no general norm for the holding of work-in-process can be proposed. This will be decided on a case by case basis.

(e) We propose no change in the existing norm of financing export receivables.

(f) A section of the industry felt that in order to make our fish products competitive, particularly in the speciality segment, and to create brand image, it is necessary to hold stocks overseas. While we do not deny its desirability we find that under the given foreign exchange regulations it is not possible. Appropriate changes should therefore, be made in the Foreign Exchange Management Act. We also find that banks will not be able to finance such an activity as they will have no control over the underlying assets. What is possible under the circumstances is for these units to open subsidiary or joint venture companies in the overseas countries. A separate scheme should be drawn up for this purpose in consultation with the EXIM Bank which may come in as an equity participant.

(g) The monitoring of bank loan and inspection of stocks have always been a ticklish problem in the seafood sector because of certain peculiarities of the industry. Evaluation and inspection of stocks maintained at the cold storages with a temperature level much below 0°C require special skills which neither the banks nor the traditional audit firms possess. We, therefore, propose the creation of an autonomous independent agency charged with wide-ranging responsibilities of stock inspection, quality control and certification, and ranking of the seafood export units in terms of creditworthiness and product quality. In fact, we envisage this body to continuously monitor the performance of the industry as an independent organization. We suggest that it should be corporatized like any other rating agency, with initial capital being provided by MPDEA, SEAI, Export Inspection Council (EIC) and banks having major involvement in the seafood industry. The name of the agency may be something like seafood Industry Monitoring and Rating Agency Limited, which shall be professionally managed under the general superintendence of the Board comprising representatives of the above promoting organizations.

For off-site monitoring, we propose a system based on cash flow projections duly agreed by the bank and the borrower.

CONCLUSION

The vagaries in the world market coupled with the ever changing food safety and environmental regulations imposed by the major fish importing countries of the world make the Indian seafood export industry more risky than any other export industry. Financial prudence suggest that a high risk industry must have adequate net worth to absorb the risk. But the Indian seafood export industry, dominated by small scale units, has a very low level of net worth compared to the risk it undertakes. Due to this, the industry could not withstand several adverse movements in international fish trade since the mid-1990s. Almost the entire industry has become sick. Nearly 50 per cent of its net worth got eroded and the real net working capital has come down to a zero level. Banks and financial institutions, having lost confidence in the industry, are not willing to render a helping hand to revive the industry. The infusion of net worth coupled with financial restructuring of the industry and installation of a suitable mechanism for monitoring and control are essential to make a turnaround of the industry. Several such measures are suggested in this chapter along with the creation of a domestic base for the seafood export industry to withstand a part of the shocks emanating from the vagaries of world market.

Bibliography

Bhattacharya, Hrishikes, *Total Management by Ratios: An Integrated Approach*, Sage Publications India (P) Ltd., New Delhi, 1995.

Bhattacharya, Hrishikes, *Banking Strategy, Credit Appraisal and Lending Decisions*, Oxford University Press, New Delhi, 1997.

Dixitulu, J.V.H., 'Addition of Monofilament Tuna Longlining System on Shrimp Trawlers and its Indicative Economics', Round Table Conference of Tuna Fishing, Visakhapatanam, 1999.

Editorial, *Seafood Export Journal*, December 1996, p. 3.

Food and Agriculture Organisation, *Fisheries Development in the 1980s*, Rome, 1984

Giudicelli, M., *Study on Deep Sea Fisheries Development in India*, Food and Agriculture Organisation, Rome, 1992.

Government of India, *The Sick Industrial Companies (Special Provisions)Act, 1985*.

Indian Express, 'Shark Cartilage', 21 August 1996.

John, Thomas A. and M. Shahul Hameed, 'Growth of Fishing Fleet of India after Independence', *Seafood Export Journal*, September, 1995, pp. 31–3.

Joseph, K.M., N. Radhakrishnan, Antony Joseph and K.P. Philip, 'Results of Demersal Fisheries Survey along East Coast of India', Bulletin of Exploratory Fish Project, 1975.

Joseph, K.M., 'Marine Fisheries Resources of India', in G.R. Kulkarni and U.K. Srivastava, *A System's Framework of the Marine Foods Industry in India*, Concept Publishing Company, New Delhi, 1985.

Korea Deep Sea Fisheries Association, *Marine Products Export Review 1996–7*, Marine Products Export Development Authority, Kochi, 1998.

—— 'Deep Sea Fisheries in Korea', April 1994.

—— *Report of the Working Group on Revalidation of the Potential Marine Fisheries Resources of EEZ of India*, Marine Products Export Development Authority, Kochi, 1990.

Ministry of Agriculture and Irrigation, Government of India, *Indian Live Stock Census, 1972* (Vol. I), New Delhi.

Ministry of Finance, Govt. of Thailand, *Foreign Trade Statistics of Thailand*, 1995.

Ministry of Finance, Republic of Korea, *Reform Plans for Improvement in Foreign Investment Environment*, 1994.

National Council of Applied Economic Research, *Export Prospects of Fish Products*, New Delhi, 1965.

Prasad, Raghu R. and P.R.S. Tampi, 'Marine Fishing Resources of India', in *Research in Animal Production*, Indian Council of Agricultural Research, New Delhi, 1982.

Reserve Bank of India, *Report of the Study Group to Frame Guidelines for Follow-up of Bank Credit*, Mumbai, 1975.

Rungarai, Tokrisna, 'Aquaculture in Thailand', *Aquaculture in Asia, Prospects of the 1990s*, APO Symposium on Aquaculture, Thailand, 1992.

Somvanshi, V.S., *Keynote Address*, Round Table Conference of Tuna Fishing, Visakhapatanam, 1999.

Supardan, Ali, 'Aquaculture Development in Indonesia', *Aquaculture in Asia*, APO Symposium in Aquaculture, Thailand, 1992.

UNIDO, *Indonesia: Industrial Growth and Diversification*, Industrial Development Review Series, 1993.

Zhiliu, Qian, *The Development of the Chinese Fisheries and Manpower in Aquaculture*, Agricultural Press, China, Beijing, 1994.

Index